On the Nature and Passage of Time
and
4-D Geometry

Samuel K. K. Blankson

WORKS BY SAMUEL K. K. BLANKSON[1]

The metaphysical Foundation for Physics;
Why time is not a natural phenomenon;
The Mathematical theory of time;
How old is the universe by our time;
The Einstein theory of space-time without mathematics;
Dead End (a novel);
Time in Science and Life---The greatest legacy of Albert Einstein;
How religious scientists play down the greatest of Einstein's achievements;
The coming revolution in physics;
Time and the Application of time;
The Logic of time in the universe;
Philosophical Essays

---

[1] These works show how my ideas evolved.

First published in Great Britain in 2012 by
PRACTICAL BOOKS

Published by Blankson Enterprises Limited
4 Poplar House, London
SE4 1NE
www.practicalbooks.org

Copyright © 2012 by Samuel K. K. Blankson

All rights reserved. No reproduction, copy or transmissions of this publication may be made without written permission. No paragraph may be reproduced, copied or transmitted save with written permission or in accordance with the provisions of the Copyright Act 1956 (as amended). Any person who does any unauthorised act in relation to this publication will be liable to criminal prosecution and civil claims for damages.

ISBN 13: 978-1-4716-8208-7 (Paperback)
ISBN 13: 978-1-4717-7551-2 (Hard Back)

A CIP catalogue record for this book is available from the British Library.

Contents

INTRODUCTION	5
THE LOGIC OF TIME IN THE UNIVERSE	26
THE NATURE OF TIME BEFORE AND AFTER EINSTEIN	70
TIME BEFORE EINSTEIN	70
TIME AFTER EINSTEIN	74
PHILOSOPHER/SCIENTISTS	80
WHAT IS SPACE-TIME	99
THE ROLE OF TIME IN HISTORY	107
PAST, PRESENT AND FUTURE	111
PROCESS AND REALITY	120
TIME AND QUANTIFIED TIME--- OR THE PASSAGE OF TIME.	123
WHAT IS MEASURED BY A CLOCK	129
THE PRINCIPLE OF MATHEMATICAL EQUIVALENCE	139
WHY SPACE ON ITS OWN IS NOT 'SPACE-TIME'	148
SOME MISCONCEPTIONS OF TIME IN RELATIVITY	156
REPLY TO SOME CRITICS ON THE WEB	167
WARPED SPACE AND CURVED SPACE-TIME	173
WHY EINSTEIN USED THE ARBITRARY MINKOWSKI FORMULA	204
HOW THEORIES OF TIME CHANGED	210
SOME FACTS, MYTHS AND LEGENDS OF TIME	214
REFUTING MINKOWSKI PROPER	241
WHAT IS ATOMIC TIME	268
CONCLUSION	271
NOTES AND REFERENCES	291

INTRODUCTION

All the theories we read about time are useful only for constructing clocks to accord accurately with the earth's motions. Don't be confused by them. The actual nature of time is unknown. However we must give the mathematicians great credit for helping to construct clocks for everyday use. The mechanics were proposed for a view of time known as absolute and cosmic time. They are still the same even after the abolition of absolute and cosmic time; and that, precisely, is the problem. Everybody still thinks about time as if it is cosmic and absolute. Well, after Einstein it is no longer so. "There is no longer a universal time." Time is not something generally existing in the cosmos of which we draw our version with our own mathematics. Yet the mechanics of the clock are still the same. This question is what I am trying to explain in this book.

The units of time (seconds, minutes, hours and so forth) constitute the time, as explained below with the concept of 'Quantified Time'. They are obtained from a breakdown of the year, or as fractions thereof. So it means they are put into the clock; the clock itself does not manufacture them. Why therefore do we assume that the clock 'measures' time for us? The actual units of time are obtained elsewhere and deliberately programmed into the clock for reproduction so that a precise number will equal to one year exactly and we start counting for another year. For there is only one year in nature; we repeat it over and over again for all the centuries.

This is extremely important in the study of time as it proves that the clock is not a metaphysical entity originally manufacturing time units for our pleasure. It is built by mechanics; and the units it produces are deliberately programmed into it for reproduction so as to coincide with a full orbit of the sun as 'one year'. Secondly, apart from the complicated theories for its mechanics, the idea that there is a theoretical version of time known only to scientists and mathematicians is absolutely untrue. Time is the same for all of us. The second is neither shorter nor longer to Nobel Prize scientists. Time's true nature is proving difficult to decipher, but that does not mean it is known but only to scientists in abstract formulations. Scientists have not the least idea of what time is and swallow arbitrary suppositions such as the one formulated by Hermann Minkowski without the ability even to evaluate them. I therefore do not subscribe to the snobbish idea that one should not write about time for the general public because the subject is too complex[2]. Everybody uses time, and it is so important that everybody must try and figure out how it came to be what it is or how we experience it. The best any writer can do is to help him or her along the right paths of logical thought.

I have found through my researches that the nature of time is discrete. It consists of discrete periods. The year and the atomic pulses of atomic time operate on the same principle, simply because there is absolutely no other method for reckoning time on earth or anywhere else in the universe. We

---

[2] In fact, the mysteries of time have come from mathematicians and philosophers contradicting one another over its true nature, added to the desire of religious leaders pretending to know what they don't know in the name of God to gain supremacy. Otherwise, primitive man just used anything to mark the passage of daylight, days and nights, the moon's phases and the seasons. These occur naturally and primitive man organised his life to accord with them. It was mathematicians, philosophers and astronomers who compounded all the mysteries about time. However the instinct of man to live in accordance with natural phenomena is still there in us, and is good enough to make everybody understand time as 'periods' created with repetitive cycles or any regular motions at all.

create or obtain these periods with repetitive cycles, pulses or oscillations like the year and atomic time. They are always 'periods' (not one period) because they are in procession---the years never end, and since other units are fractions of the year, time is virtually continuous. That is the solution of the passage of time. Yet it is secular. It is secular because it is not the real time (whatever it is) but just how it is passing by only---we count the mere physical orbits as years. The real nature of time is unknown. I do not accept that time does not exist but that it can never be defined other than using repetitive cycles to show how it is passing by.

This definition of time makes it seem simple; but of course it is simple. We know we are orbiting the sun and call each orbit 'a year' and pare the year down to all the other units of time, including atomic time since it has to be based on the second to make sense. However, due to the mysterious nature of time, how we experience it as something very closely linked to life and how human beings have mythologised it over so many thousands of years, this definition requires patience and hard work to make it conducive to the human mind, and I have been working on the problem for more than fifty years. It does not only have to be true; it has to be acceptable to the human way of thinking about time. Hence, in what follows, I have had to base my explanation of this simple theory of time on eight rather complicated postulates. You may say perhaps I am trying to emulate past great thinkers like Euclid or Einstein. I am afraid that is not the case. I just find the whole idea of time so messed up with myths that these postulates are required to make things a little bit clearer. Since the postulates will be familiar to readers, how they are discussed here may also be found interesting:

1. That because of relativity there is no longer a universal time---Russell (ABC of Relativity, Ch. 5).[3] His further question

---

[3] The full statement is this: "There is no longer a universal time which can be applied without ambiguity to any part of the universe; there are only the various 'proper' times of the various bodies in the universe." Einstein put the same idea thus: "There are as many times as there are inertial frames. That is the gist of the June Paper's kinematic sections..." (Abraham Pais, p141, see below). The obvious implication (or logically necessary inference) is that time is limited to a frame, since Einstein said he got the idea from the Lorentz local time concept. Those who have analysed the matter have come

regarding what is [therefore] measured by the clock is answered in the text below.

2. Due to the theory of 'Frames' every 'body' has to have its own time so that there are as many times as there are bodies---Einstein (the kinematic sections of the June Paper[4]);

3. That the all-embracing time is a 'construction' like the all-embracing space; physics itself has become aware of this through the discussions connected with relativity---Russell (Mysticism and Logic, Ch. viii(x));[5]

4. That time is unknown; we use repetitive cycles, like the year, to show how it is passing only, therefore the arrow of time idea is not needed. The years replicate; they need no arrows to make them do so;

5. To have what we call time sentience is required---somebody must be there to count the years otherwise how could we get the centuries and the very concept of ageing?

6. That Hermann Minkowski (see below in the notes)[6] relied on imaginary time coordinates therefore his equation of space to

---

to the conclusion that time comes from regular or repetitive motions in a perspective (sentience is required); therefore what we know as time is how it is passing through nature only---the years, for instance. Atomic time is the same since it is based on regular pulses or oscillations. It is the regular cycles we count as the rates of passing time. The real nature of time is unknown---e.g. the year is our basic unit of time out of which all other units are derived, but the year is merely a physical journey round the sun. It is man who counts the orbits of the sun as the basic rates of the passage of time, in the same way as we count the oscillations of atoms as another way of knowing how time is passing and call it atomic time.

[4] If in doubt consult Abraham Pais's excellent Biography, 'Subtle is the Lord...', Oxford, 1982.
[5] I must remind the reader that we are talking about time, that mysterious thing which makes people age and previously said to have begun at the time of Creation. Even scientists call something 'Time Zero', as the logical beginning of a universal time---so, obviously, they are mistaken. When they realise their mistake a lot in science is going to change.
[6] He first burst on to the scene with a speech about space and time, usually mentioned in German as *Raum und Zeit*, in Cologne, November 1907. He followed this with a full technical paper, see Goett. Nachr, 1908, p. 53. Reprinted in full in *Gesammelte Abhandlungen von Herman Minkowski*, Vol. 2, p352, Teubner. Leipzig, 1911. His only fault is that it is impossible to

time is logically invalid and his concept of space-time is wrong. In other words, there is no such thing as 4-D geometry or four-dimensional continuum incorporating time into space so that time and space constitute one entity. 'Space-time' must be interpreted as meaning that time is obtained from space because every unit of time is 'relation between points', as Russell put it---the year, for instance. Thus Minkowski is completely irrelevant.

7. If there is no longer a universal time then there is no time in any part of the universe where there is no sentience, and applying earth time there is wrong because it is in breach of the Einstein theory of 'Frames'. Somebody must be there to set the points and count the years, etc. This is the really shocking new idea about time that is going to take mankind ages to understand; when and if we do, quantum physics and even physics itself will be seen differently. For there is no such thing as the general passage of existence, since existence is individual and never general---it involves billions of things all passing through nature differently. There is vegetation and animal life but they are not general; it is not general existence; every 'thing' or animal is different. General existence implies instantaneous but multiple generation which is unknown on this planet. It may occur but only accidentally Things and animals come and go individualistically. All existence on this planet is individual. It is doubtful if two people see one thing identically; even if they do there is the time element.[7] The time coordinate between one person and another makes them absolutely distinct individuals---with two different perspectives.

8. The whole concept of earth time is based on the 'Day&Night' system: it is the source of planning ideas (i.e. knowing that daylight and night time would provide hours for doing whatever we want to do), and notions of past, present and future; also our inherited instincts about moving forward, ageing as the days pass by, and the passage of time as a whole. The

---

equate space to time except with imaginary quantities which are unacceptable. Russell said it is compounded for the convenience of mathematicians without criticism, but if physical reality is not four-dimensional then what is the point? Physics is not a joke, is it?

[7] If in doubt we can invoke Einstein. The analysis of order and simultaneity shows that two people cannot see one thing identically. The general passage of existence through billions of individual perspectives? That is fiction.

revolutions of the earth gave primitive man the sense of time by counting the days. Modern man added theory, but counting cycles is still the basis of our sophisticated time.

I must explain with an apology that this book was originally planned as a book of disparate "Philosophical Essays—Mark II", since my other book called "Philosophical Essays" did not strike me as satisfactory. So, as a treatise, the book is not properly planned as a coherent unit since I changed it from one type to another in the process. My plea is that writing about time is torture enough and beg the reader to be charitable!

Now, in my most optimistic private moments where I could not be accused of vanity, I say to myself that the time has come to lay the Minkowski equation of space to time to rest in peace like its creator. At the same time I fear that any publisher interested in my ideas would have to take them as they are without the necessary academic embellishment as would be found in academic text books because my subject, time, is not only difficult but deeply mysterious, even scary.

It is foolhardy to try and change mankind's established knowledge of time---though that is what I am trying to do.[8] But so difficult is the subject that, try as I may, I cannot write my ideas out as smoothly as I would have liked. Part of my problem as mentioned above is that the book was originally planned as a small book of essays mostly culled from my published works. But one thing led to another till I found myself writing a mini technical tome about time. Given the importance of the topic discussed here, I hope the reader will agree with me that the presentation, however unsatisfactory due to the difficult nature of the subject, is really a small price to pay for such a theory. In the end, all theories are mere suggestions put forward tremulously for consideration by the reading public; but the writer also deserves a little consideration for his courage!

Thus this book has become semi-technical. There are only two ways of writing technical books. You either write for the

---

[8] Somebody had to try, for the present situation is unacceptable: we can't tell whether time is the same thing as space or not. The mathematicians say it is but they cannot prove it except with imaginary quantities in mathematics---which, of course, is unacceptable---while some philosophers are telling us that there is no time at all in the world. (See A World Without Time: The Forgotten Legacy of Gödel and Einstein, Penguin, 2007.)

professionals (or the academics) in a very opaque technical language or broaden the text to appeal to the general public. But, generally speaking, philosophy alone is for philosophers. The subjects discussed are so intricate and abstract there seems no way of making it accessible to the general reader. However I take the view that, although time is assumed to be a philosophical subject, everybody uses time so it must be presented in a form that general readers can understand.

Strictly speaking, the theory should have been aired in a technical journal first, but no journal will touch any paper with negative ideas about Hermann Minkowski, the great mathematicians who discovered 4-D geometry and made relativity accessible to scientists, except that, in actual fact, he didn't do any such thing, or he couldn't achieve it successfully. Hence I choked with laugher when the CERN announced publicly on television that they're going to re-examine the theory of 4-D geometry because they thought c has been breached. The point is, you don't get the taxpayer to spend so many billions of dollars on a project if you are not even sure of your theory. Minkowski is not wrong because of c. He is wrong because of the i in his equation of space to time; and Eddington, Whitehead, Russell, etc., told us so years ago.

In a way, one has to feel sorry for physicists. After the propagation of light, the cause of gravity, relativity, the Quantum theory and QED, there is not much left for them to do, hence the CERN projects, most of which is sheer waste of money. As I see it, time could have provided a fertile ground for much useful work but the Minkowski formula is blocking the way---it may be blocking the way for the Theory of Everything, too, and they had better take a good look at it. It is interesting that when the CERN thought they had breached c, they agreed to re-examine the concept of 4-D geometry to see if it is true. Well, be that as it may, Eddington told the world that it is fictitious nearly a hundred years ago. Now we are told that Einstein too did not accept or even understand it. When you say that Einstein did not understand something in physics, you are saying, in effect, that he thought it was nonsense.

For good or ill, I have addressed the book to both the professionals and the general readers among the population interested in the nature and passage of time or what we refer to as time.[9] The rest of mankind take time for granted, or think it is too difficult to interpret. However, I am humbly telling everybody that once the years are recognised as the passage of time, the inferences based on that notion (or fact) about time are all reasonably easy to understand. The logical reason is that the yearly cycle is ordinarily physical. So it means we use mere physical cycles and count them as the rates of the passage of time when we regard the passing years as the passage of time---simple! For every unit of time is derived from the year as fraction thereof. But standing in the way is the Minkowski theory of space-time, which, I suggest, must now be laid to rest in peace because it is logically invalid for being based on imaginary time coordinates.

As I said, the CERN (now known as the European Laboratory for Particle Physics) has admitted publicly that because they think some particles can travel faster than light, the equation of space to time may be logically invalid.[10] Since I have already published ten books denouncing the equation of space to time, this book is also addressed to the CERN to welcome them to real physical reality; but I fear that once the Minkowski equation of space to time is rejected, a lot of theories in physics are going to need amendment; therefore scientists will not pursue this matter too vigorously. The CERN is changing their minds in panic again. Yet if it is true, as we are told by great mathematicians (Russell, Whitehead, Eddington and even Einstein himself) that the Minkowski formula is arbitrary, then I prophesy that the CERN will encounter it in panic again and again because time is crucial in physics and chemistry---it is a catalyst. I used to believe that in theoretical physics mistakes do not hurt anybody but time is different; besides, Minkowski is using false time for the determination of the nature of physical reality. Mathematics can be used to iron out the discrepancies in time; but physical reality is not, and I repeat not, a four-

---

[9] It is addressed to everybody who can read because it is humbly and tentatively suggesting a theory of time that, if successful, could settle the argument about the nature and passage of time for good---it has to be aimed at everybody because we all use time and want to know what it is.

[10] As I write they have come back to announce that c has not been breached after all; but that does not mean the Minkowski equation of space to time is correct. The scientific community, especially in astronomy and particle physics, are urging us to spend several billions on their projects, yet they're not even sure of their basic theories.

dimensional continuum. The Minkowski formula is not scientific or logically valid because it is based on imaginary time coordinates. Thus since all physics at the moment is built on the supposition that space is four-dimensional, CERN is bound to come face to face with the fact that it is not---and I predict that it won't be long. At present nobody will listen to me. But it won't be long that some dishonest professors will write a tome stealing my ideas to get the Minkowski formula abolished---and be honoured for it.

Yet I have to stress that my theory is tentative for that is the correct and humble manner in which theories about any subject as serious, fundamental and mysterious as time, ought to be introduced. Although I believe the theory is true, no man should play God among his fellow men but always acknowledge that they have as much rights as everybody else. Suppose a thinker finds the true meaning of life and wants people to do something because of it, he'd still have to go about it in a charming, humble manner otherwise they'll throw him out. The situation is the same with time.

Time and life are the two most serious subjects we want to understand. I put the quest for the meaning or origin and purpose of life first, time second. As such, no human being, not even an Einstein, can present his theory about them as definitive; rather it must be humbly presented as his own fallible, human and tentative suggestion for consideration by all of us, no doubt with underlying logic in an attempt to convince us. It would be the height of irresponsibility, arrogance and contemptible vanity for anybody to claim that his theory is true. Nobody knows what is the truth about time because nobody can even know what it is. All through history what we have been calling time is only the rates of how it is passing by---which we get by the use of regular or repetitive cycles. People who claim that time travel is possible should please tell us what it is first.

Concerning this matter, I am worried that science is losing the battle with religion, which is on the ascendancy again and many books are coming on the literary scene about the possibility of time travel, a religious idea closely related to life after death, namely the belief that life came from somewhere and shall return to it after death; and that time travel forwards and backwards is 'a scientific possibility' and therefore death is not the end of life.[11]

Of course, we all want to come back after death, but is it possible? The Minkowski equation of space to time coupled with the concept of curved space are assumed to be evidence that time travel is possible. I agree that the idea would be worth considering seriously in science if the Minkowski theory were true in actual reality. Unfortunately, it is only true in his mathematics, but clearly false in nature. Mathematicians promote it in the use of the term 'space-time' because it makes things simple for them---although they know very well that it is false. I keep wondering how long they will go on trying to fool nature with that false theory. For one thing we know is that physical reality tolerates no falsehoods. If the moon was only a mirage, we couldn't go there. So the theories that it is substantial had to be true. Nature accepts no falsehoods of what it is.

Concerning the book, still several of the essays here have been selected from published works, and this long Introduction has been designed to explain the book well so that readers can get the gist of my argument at a glance. It is in this Introduction that technicalities of the subject matter are minimised to help the general reader's understanding.

It is a good advice to the general public not to read a philosophical discourse direct without interpretations and try to understand it. Usually there is mental torture which may be completely unnecessary.[12] They are present in some philosophical pieces, especially from thinkers like Hegel, Kant

---

[11] I am worried because religion causes wars.

[12] Given what Sir Karl Popper has said about the kind of philosophers mentioned above, viz: "You see, the history of man is a queer thing. It's a history of a succession of attacks of intellectual madness, of all sorts of strange intellectual fashions. I don't need to give many examples of revolts against reason...for we know how strongly certain fashions have taken hold, not only just in a comparatively small insular group, but, in large parts of mankind. Russell saw these things in that light, and so do I...In the long history of philosophy there are many more philosophical arguments of which I feel ashamed than philosophical arguments of which I am proud. Yes, I cannot say I am proud of being called a philosopher."(Quoted by Bryan Magee, in his book *Modern British Philosophy*, London, Secker & Warburg, 1971.)

and Wittgenstein, particularly the latter, who could reduce Bertrand Russell to tears with his 'logical mysticism'.[13]

On the other hand, it is hard work, yet I aim my work at the general reader for two reasons: one is that the academics ignore me, even though they steal my ideas. I have conclusive evidence of that. The second is that I am discussing time in a format known to everybody---second, second, second; or year after year after year. We all know that the years and seconds are discrete. By thinking hard about the reasons I discovered that time cannot be had in any other way, which suggests that we know how time is passing only.

This means our time is discrete; and I argue that discrete time needs no arrows to make it pass by, and so forth. Surely, everybody can understand this in simple books? There is no need for hefty tomes, and even less need for opacity, which is often used to conceal discordant ideas.

I begin by pointing out that a little over one hundred years ago, intellectually everything in the world changed (without any fanfare)[14], when the Dutch physicist H. A. Lorentz and Albert

---

[13] I have often wondered why in Heaven's name Russell wasted so much time and energy to entertain Wittgenstein. I believe it is because Russell, being one of the most humane great philosophers (who wrote to me when he was 90, and I was just 22---and signed the letter himself), was a very kind man indeed. But consider this: Russell concluded on firm logical grounds that physics is to be obeyed on pain of death (My Philosophical Development, Ch.11); and yet what he described as Wittgenstein's 'logical mysticism' "puts an end to physics" (Analysis of Matter, Ch 2); even worse, physicists have nothing but contempt for all philosophies that question their discipline. So why bother with somebody who wants to abolish what cannot be abolished on pain of death? Those academics (especially in Oxford and Cambridge) who consider Wittgenstein as the greatest philosopher of the 20th century are not real philosophers, the sort that give philosophy a bad name--- as Popper is saying.

[14] Religion ruled the world in those days and all notions of time were, officially, religious. Man's scientific ideas were God's merciful endowments. After all, God created the world 'precisely' at noon in 4004 BC, apparently using clay for human muscles, and the date 4004 BC, too, apparently, had been running from 'Time Zero' to the day of creation. The reader can see how the scientific notion of Time Zero was borrowed from the religions! I have absolutely no respect for religious scientists. They should have chosen different professions. Science is for scientists----i.e. logical thinkers to whom superstitious concepts from the religions are anathema,

Einstein discovered that time is neither general nor absolute, and, therefore, one second here is not one second everywhere else. The suggestion that there must be differences about how sentient beings in the cosmos get their seconds (or units of time) is monumental and totally without equal in the history of human thought simply because time is the most important thing in life except the life itself. Of course the new theory of time is not that simple; but how we get our 'different' units of time in the cosmos is the core of it; and that, essentially, is what makes Einstein a great philosopher as well as scientist of genius—extraordinary genius, to be exact. It helps to refute the notion of absolute or general time running all through the cosmos and the same everywhere. Given the mysterious nature of time, the religions were right: immutability made the old theory of time into something created from outside the cosmos and bestowed upon it to run all through it in exactly one format. It thus became the last hiding place of God after Darwin. For once the secular way we get our seconds is understood, cosmic time is abolished. This is where Bertrand Russell, our most recent great philosopher, came in with his question as to what then is measured by the clock, as discussed below in one of the essays.

Yet time seems natural, extremely mysterious and the second most important thing in life. How did we create something that is natural; and yet, otherwise, where did our own unique seconds come from since Einstein and Lorentz have solid experimental evidence to support their theory?[15] This has created the greatest conundrum in human life, because it means we have to try to establish the logical structures of our time and how it began. For fifty years I have devoted my life to working for a solution to this quandary; and I am going to share my ideas with the reader in these pages.

---

something like night and day: they just cannot meet.

[15] This is how strongly Professor Eddington put the case for Einstein:"Prior to Einstein's researches no doubt was entertained that there existed 'true even-flowing time' which was unique and universal...Those who still insist on the existence of a unique 'true time' generally rely on the possibility that the resources of experiment are not yet exhausted and that someday a discriminating test may be found. But the off-chance that a future generation may discover a significance in our utterances is scarcely an excuse for making meaningless noises" (*The Mathematical Theory of Relativity*, Ch.1.1.)

Since the days of Lorentz and Einstein, philosophers have argued that what we call 'time' is not time at all but the rates of motions used as the rates of passing time, and that the true nature of time is probably beyond the comprehension of humankind. Indeed, as the great Bertrand Russell has pointed out, under relativity the 'all-embracing time is a construction', a human artefact. But he made it as a comment without deep analysis. Here is my analysis: if there is no longer a universal time (his own phrase originally given in his ABC of Relativity), and the time we have is 'constructed', and can only be based on points---as the earth-year shows—then our time is necessarily discrete. Discrete time needs no arrows to pass by. Rather, the successive units cause the passage of time, as the example of the year shows by replicating to become the centuries, and so forth. It is not difficult to understand that our time is discrete because the year out of which all other units are derived is determinate, and has to be repeated for time to continue; its fractions, too, which are the other units of time (from the second to the months and so forth) are also obtained with points or mathematics and therefore also discrete.

The method we use for this amounts to quantifying time whatever it is---the pulses or oscillations of atoms in atomic time are logically not different from the orbits of the sun.[16] In other words, we use mere physically repetitive motions or cycles, and call each cycle, say, 'one year' as we do. This is to quantify time, to reduce it to units of duration or periodicities that we can use for cultural purposes in conjunction with the essential features of the world.

The most prominent of these motions (or regular cycles) used to reckon time is the earth year, which is not time at all, but the physical orbit of the sun by the earth. Yet we regard the passing years as the passage of time, simply because we know nothing else as 'time'---or to use for time. Let me illustrate this with language. We normally say "after the lapse of time", say, after the lapse of ten years. But in reality it means after ten years (ten orbits of the sun) have passed---so what we call

---

[16] My guess is that time is the product of motion, physio/chemistry (being the physical causes of change), sentience, the ability to count and the intellectual use of points—added to the sheer accident of existence through the collocations of atoms. However, anybody who can make sense of this is a better thinker than I am!

'time' (or ten years), is also the passage of time; and yet that is all we know as time.[17] My argument is that it is not only what we call or know as time but also the only thing we can ever know of time. We cannot reckon time in any other manner---not only on this planet but throughout the universe. Whatever time is, it is certainly universal: as cosmic, it is universal; conversely, as discrete time that is what it will be throughout the cosmos. The logic is straightforward. If there is time in the traditional sense, of course it will cover the whole universe; that is why we thought there was universal time. On the other hand, if we have found that there is no such time, the new condition will also cover the whole universe, for we are part of it. And whatever we do to get what we call time is also the only method available to any 'Beings' in this same universe. Taken in this sense the book is not difficult, and I think intelligent readers, not only academics, will enjoy it. However the thesis has implications for cosmological studies of ETs, and I hope cosmologists will take notice of it.

Atomic time is often mentioned; but I have explained in this book that it is not different from the yearly cycle; logically the principle is the same, rather the atomic time units are minute and therefore less convenient, culturally, than the earth-year, sub-divided down to the seconds and so forth.[18] The greatest advantage of this interpretation of time is that it solves completely the ancient problems of the passage and continuity of time. For, alas, it seems that all we can ever know is the mechanism of how time is passing (which makes it seem

---

[17] We use the orbits of the sun as our knowledge of the rate of the passage of time.

[18] Atomic time is defined as "a technique that allows atomic transmissions to be measured with great accuracy." Two points need to be stressed here. (1) The word 'transmissions' (that is, in plural) means the atomic pulses or oscillations are regular and therefore not different from the orbits of the sun, only shorter in duration; and (2) the phrase 'great accuracy' means it is based on or compared to something, namely the second. Thus both atomic time and earth-time merely show how much of time is passing and never the true nature of time. Atomic time cannot be used to contradict my theory. There is always religion in people's discussion of time. It is so mysterious that they tend to regard it as divine. Exaggerated claims for atomic time are a case in point. Atomic time is used to deepen the mystery about time. In fact, it is the same as earth time---it can only show how time is passing.

continuous) and never the true metaphysical nature of time. This, in a nutshell, is the subject matter of this book.

As mentioned above, many of the essays have already appeared in other publications of mine either as they are or have been redacted; in some cases even completely rewritten. They overlap; also, there are several repetitions; but the reader should pardon me because it is a tough job trying to change notions of such a basic thing as time without repetitions for emphasis.

Everybody thinks he or she knows what is time; yet technically (in reality), nobody knows what time is either in science, philosophy or religion. Like life, we use it, but we do not understand it. Nobody knows what it is. Time is so mysterious that it is not possible to discuss it in simplistic terms and expect people, even philosophers, to accept new theories of it on trust or even authority---and even then what or whose authority? Even the Popes fare no better.[19]

Rather the writer must turn it this way and that, giving several explanations of each facet of time and generally labour the points for people to take notice out of sheer perseverance (confusing mere perspiration with genius). It still will not earn the writer any accolades but at least people will take pity on him for his efforts—although the mathematicians will still sneer because they think, wrongly, that the world consists wholly of the Minkowski space, which they sometimes call Minkowski universe, or 4-D geometry. I will leave it at that so as to avoid unnecessary technicalities that may bore the general reader.

These pieces have been selected and specially revised (in part) with due care to make them easy to understand, and it is hoped that the reader will find them reasonably clear and unambiguous. The fact that they have been published before should not matter too much if the aim of the reader is to discover truth, the main purpose of philosophy.

As for motive, my chief aim for writing this book is to whittle away some of the mathematical, religious and philosophical

---

[19] Albert Einstein alone (in all history) succeeded brilliantly by simply showing that it is different in different places---so it cannot be either absolute or general---and therefore set us the task of discovering how our own time began, and so forth.

complexities surrounding time, or the nature of time, as we know it on this planet, so as to bring it within the intellectual compass of everybody who can read, including those mathematicians who seem incapable of living in this world without thinking of it as a mathematical space in the mythical four-dimensional geometry of Hermann Minkowski---that the mathematical space of space-time has to have an imaginary time coordinate to exist and yet the end result is time known as 'space-time' fails to excite their critically faculties.[20] An imaginary time coordinate has no place in 'the logic of time', since everything on earth has to have logical credentials---see below for my essay about 'The logic of time in the universe'. What I am saying in "The Logic of Time" is this: if it is true that under relativity there is no longer a universal time then how can there be time in the cosmos at large outside the earth? Or, nearer home, how can we have time in the metric of general relativity? Cosmologists are applying earth-time everywhere, but that is a logical anomaly. There is no time in the cosmos. Time can only be created by sentient beings. The fact that these cosmologists and astronomers ignore my work (whilst some of them are stealing my ideas about the impossibility of deciphering the secrets of the universe with man's limited brains) does not make them right.

Mathematicians (always) get away with quite a lot. I don't know why we tolerate their practice of inventing rules to suit their nostrums. Many of them have told me from exalted positions in the journals that the Minkowski imaginary time coordinate is just to make the metric suitable for the supposition that space is four-dimensional: "...the role of i is just to make the metric come out right..." Therefore by criticising Minkowski my papers should be rejected.

---

[20] One popular Encyclopaedia called space-time an artificial concept. Therefore, myself, Bertrand Russell and Professor Sir Arthur Eddington, and one or two others, are not the only thinkers who regard the Minkowski space-time as unworthy of science: "Spacetime (sic) is undoubtedly an artificial concept" (Routledge, *Concise Encyclopaedia of Philosophy*.) It is---artificial and bogus as a science---yet mathematicians insist that it has successfully linked space to time in 4-D geometry. In any academic discipline, it would be condemned except in mathematics. Mathematicians rather condemn anybody who is too ignorant to realise that Minkowski has linked space to time with mathematics therefore 's=ct...'

Let us look at this idea in a little detail---the book contains most of my detailed objections. But now, let us consider this: physical reality is three dimensional. To add time we use the 3+1 formula. Then Minkowski came along with his theory of 'Space-time', saying he could show that the whole metric is, in fact, four dimensional, meaning that time is incorporated into space. It made things easy for the mathematicians (as Russell noted in The Analysis of Matter, Ch. XXXVI): "SPACE TIME, as it appears in mathematical physics, is obviously an artefact, i.e. a structure in which materials found in the world are compounded in such a manner as to be convenient for the mathematician." So it is not true in physical reality; it is an artefact---besides it is based on imaginary time coordinates. Professor Eddington, a great scientist, called it 'arbitrary and fictitious'. But the point is that everywhere we hear of space-time. It dominates all physics, cosmology and part of astronomy for the convenience of mathematicians. To my mind, this is a distortion of relativity and physics as a whole; yet the mathematicians are not concerned. All mathematicians believe that mathematics can be used to alter physical reality; but I argue in one of the essays below that this is not the case.

As Professor Sir Arthur Eddington and the mathematical genius Paul Dirac have pointed out, mathematicians themselves invent the rules in mathematics specifically to suit themselves as against all others. According to Professor Eddington, "The pure mathematician deals with ideal quantities defined as having the properties which he deliberately assigns to them." However, the very strong caveat against this almost dishonest intellectual practice was stated not by philosophers but by Professor Eddington himself. He wrote, "But in an experimental science we have to discover properties not to assign them."[21] Going by this rule, and since Eddington famously described the Minkowski formula as 'arbitrary and fictitious', I contend that it

---

[21] From the Introduction to his, *Mathematical Theory of Relativity*. You never get the impression from this book that Professor Eddington's criticism is aimed at the Minkowski theory of space-time; yet it was. Nothing in Einstein's ideas was either arbitrary or fictitious. Eddington treated the Minkowski theory as if it is textbook science; so if he says it is arbitrary and fictitious then his own exalted tome based on it is flawed. The scientific establishment is in trouble because it is going to be almost impossible to stop calling time 'space-time', even though the Minkowski formula is false.

is wrong for physicists to use it for the serious business of determining the nature of physical reality---namely, as to whether or not it is four-dimensional. My technical papers have been ignored for nearly fifty years because of my attitude; I am dead against the Minkowski formula for equating space to time in his 4-D geometry.[22] So it is up to the reader to judge whether I am right or not. My own opinion is that mathematicians are using the Minkowski fiction to distort physics and philosophy. Unfortunately, it happens to be the most difficult subject on earth; as a result, not many thinkers can understand it let alone have the courage to question the Minkowski theory about it---i.e. as to whether or not physical reality is four-dimensional.

Some of the merits of this novel interpretation of time are that we can infer that time in every part of the universe will also be 'a construction' and roughly similar to time on earth, which is what I call "The logic of time in the universe". It makes time necessarily discrete so that history is seen as the march of events and not the march of time whose units (and therefore its essence, like the years), are literally just passing by. Furthermore, the concept of curved space-time and the 'Time Zero' idea based on it cannot be logically sustained---and therefore time travel is impossible since the Minkowski equation of space to time is indefensibly flawed, even Professor Sir Arthur Eddington and Bertrand Russell said so. I like to repeat this in case mathematicians have been rendered dumb by the groundless equations of Minkowski.

Above all, let me stress again that discrete time, known only by how it is passing by (since the passing years is passing time), does not require any theories to explain how it passes by and seems continuous: the passage of time is the passage of the years; therefore time passes by through the replications of its units---the years, for instance---and the arrow of time idea is no longer needed. The procession of the units of time (like the successive years) causes the passage of time, and so long as the succession or replication continues year after year after year, there will be perpetual time on earth. However, looking far

---

[22] I fail to see how it can be regarded as part of physics---as scientists do now. Each time they mention the term 'space-time' they are implying that space and time are merged by means of the Minkowski arbitrary formula.

ahead, when the sun dies our time will die with it. The mechanisms for constructing our time will disappear.

The legends of time are legion; every tribe on this planet has its own beliefs about time, most of them illogical. Scientists are a tribe that is supposed to live by rational thought and so they also have their own legends about time, some of which are more conducive to rational thought than others. To my mind, the scientific notions of time that are misconceived derive from the Minkowski equation of space to time. So I have discussed some of the topics I consider important (such as Time Dilation, Clocks Paradox, Twins Paradox, Curved Space-time, Gravity and time, etc.,) from the point of view of logical thought and not the Minkowski theory of 4-D geometry---which, I repeat, is logically untenable.

As always, my most sincere thanks go to my son and publisher, the prolific writer and computer buff, Samuel Blankson, for his help as my editor, agent and publisher. Alas, we work alone. Nobody wants to look at my work. In addition, nobody wants to see or read the evidence against the Minkowski formula for 4-D geometry. For this evidence has been stated in more erudite books by Professor Sir Arthur Eddington, the founder of Astrophysics, Professor A.N. Whitehead and our great Bertrand Russell over a century ago. Nobody is interested. Some have mentioned 'mathematical refutation'. I don't know what they mean, for the point is, the theory is based on flawed logic; so the subsequent mathematical deductions are all flawed.

The reality of the matter as regards time and the Minkowski theory of four-dimensional continuum is that his formula very strictly tied man's mind up so severely that the only way to free mankind's intellectual creativity was to scrutinise the logical foundation of his formula in the hope that he might be wrong; and when we did so found it defective in logic not in his so-called beautiful mathematics. The mathematical deductions are correct so long as i is allowed as valid to represent time; but it meant that time was imposed because it was inherent in every space. So it is not just a matter of mathematical refutation; it is sheer humbug to talk of mathematics when the logical basis is untenable. As always, I believe Einstein was absolutely right to pretend not to understand the theory. For when an intellectual giant of the calibre of Albert Einstein described any theory as such, as far as I am concerned, it means he thought it was nonsense.

The three great thinkers mentioned above were all great mathematicians; yet even they did not attempt to rescue or refute the theory with mathematics. They just said it is arbitrary and fictitious. It is true that Eddington urged mathematicians to continue to use it; but that was because, at the time, it was the best way to understand relativity. Then the philosophers came in to say, 'but it is not true', therefore it is unacceptable as the means for the interpretation of physical reality. By this judgement physical reality reverted to the 3+1 formula as the true state of physical reality. Of course in mathematics we often wish things could be different to make life easy for the mathematician; but that is wishful thinking, and wrong to assign properties, as Eddington put it, to make it appear that it is the natural state of the universe when it is not so in fact. Unfortunately everybody knows that that is what the Minkowski formula for equating space to time amounts to, yet mathematicians continue to cling to it.

Atomic time is said to be different from the time we get from orbits of the sun; in fact, it is the same; I am stressing this because scientists claim it is another 'different' way of measuring time, implying that there is universal time measured in many different ways by different methods. Scientists make mistakes about that and many other things; they should respect philosophers for pointing out the obvious errors that regularly occur in their philosophical interpretations of nature. Establishing a fact is one thing; interpreting that fact logically, or any fact, even in works of genius, to fit into the general, logical (and epistemological) scheme of nature is quite a different proposition. The pulses or oscillations of atoms are not different from the repetitive cycles (of the years) to show how much time is passing by (they are merely shorter cycles), which is all we can ever know of time. Historically many objects have been used to track time, but they are all the same---using repetitive cycles and counting them as the rates of the passage of time.

This interpretation of what we call time makes it seem simple; instead of being seen as a mysterious conundrum without knowing whence it comes, we just have to look at how we get the years and the necessary inference becomes clear, namely a procession of time units replicating all the way to the centuries. However it is analysed, we will find that time (in all history and everywhere) is always reckoned with passing units of it and never what it really is---no wonder many thinkers infer that it does not exist. The problem is that there is something we use to regulate life on earth and we call it 'time'. What I have found is that it consists of regular cycles. And all through the book I have treated atomic time as conceptually similar to earth time. I hope the reader agrees with me.

KWASI BLANKSON
London, 2012

THE LOGIC OF TIME IN THE UNIVERSE

Under relativity there is no longer a universal time, as confirmed with one of the postulates taken from Bertrand Russell no less. Thus we have to examine how we get our time. It is argued here that there is a method or logic involved, and that it will even apply to every 'Being' in the universe. I conclude that in any case nobody can ever know the nature of time, because what we think is time is, in fact, just how it is passing only; so we can only ever trace how time is passing and never what it is. The logic of time just shows us the passage of time; therefore there is no need for any other theoretical explanation for the passage of time.

Let me begin with a quandary the solution of which can win somebody the Nobel Prize. The eternal question "What is life?" is, on closer examination, associated with or inextricably linked with the question "What is time?" As evidence, we are reminded that we begin to age the moment we are born, and ageing is associated with the passage of time. Is this ageing 'time moving you on' or 'time moving on' independently? In either case life and time seem identical in as much as they are the only two basic things in the world of sentient beings that can never be logically defined---and one is dependent on the other, or cannot be separated from it. I find this situation rather intriguing, philosophically challenging and scientifically insoluble. My speculation is that whoever solves the problem of time would make it easy for us to understand the nature of life, if not where it came from, or why it came to be at all. Since we die what is the purpose of life? The most logical supposition is that it is a chemical mistake. The same way that a chemical accident resulted in the creation of Mr Hyde. I believe that story is really profound.

In the mean time, it is worth pointing out that living is using time to age; ageing and living are inseparable. So is it the same thing as time since it is in the consciousness of the living the time is noted? Secondly, we cannot define life and we cannot define time and yet the two go together inseparably. For an individual the end of time is the end of his or her life; and the beginning of time is the moment of birth. Therefore we need to investigate seriously the suggestion that life is time and time is life that is why nobody can define time and yet it is making him or her age incessantly.

I have some doubts, though, about all this, most of which come from optimistic, jackboot scientists. Ageing can be seen as mere chemistry;[23] also it does not necessarily move forward like time. If an infant is not given due care it will die soon after birth. So life seems to me to have a slight edge over time in that we can make time secondary so as to appear independent of life by using repetitive cycles to demonstrate how it is passing through nature as a separate entity from life---the years, for instance. And this is, for me, part of the logic of time in the universe. Time appears to me to be ultimately chemical in nature. And we are accustomed to these chemical processings. We are aware of the chemical nature of life and accept it for there is nothing else we can do. We eat food and

---

[23] Chemistry causes waiting, and waiting is time moving on. The point is we appear to be basically chemical products so waiting for chemistry to do its work is inevitable. The waiting is chemical, but the time or the configuration of units of time is derived from repetitive cycles---not as mysterious as we suppose because we ourselves create it to guide our activities. It is artificial. In the cosmos at large there is no time. "There is no longer a universal time". The genius of Albert Einstein was much more wonderful than we have so far realised.

wait for it to digest and give us energy; take medicine and wait for it to take affect; light a fire and wait for the radiation of its heat; put food on the fire and wait for it to be done and become eatable---and so forth. In everything we do waiting of some kind is involved, and mostly we know that it is a chemical process. They do not constitute time *per se*; they are chemical processings. They do not also move life forward, not necessarily; some foods will kill you. However, they all involve waiting---that is time for us. That thing which we call time and what we do with it intellectually is all human in origin, not part of the elements in the universe. In the universe they are mere chemical processes.

Furthermore, time seems to have evolved late in human existence. The sophisticated clock took centuries to perfect. You can also live without time---blind people do. Life, then, is obviously not identical with time. Life came first; time evolved later. Solving the problem of how it passes and seems continuous is miles away from the nature and purpose of life. In a 'little' digression, I would like to comment on the Einstein division of the cosmos into the two definite metaphysical categories of (a) where life is possible (in a special relativity frame); and (b) where there is no where anywhere (stable) for life to evolve and flourish---i.e. in the metric of general relativity.

The wonder is how he conceived the notion. The law of inertia and Riemannian geometry probably helped him. Yet that is pure speculation, which, even if true, shows his incomparable genius as a logical thinker in all history. Aristotle was almost the same, but he was full of errors of judgement, of inferences and further consumed by primitive religious ideas of the Hellenic age. We'll never know whether he was afraid to contradict them, because he was also some brain, most definitely on a par with that of Einstein. Man has almost always got things wrong so that at the time of Einstein there were numerous theories tearing science here and there.[24] For one man to correct many of these and add the quantum theory and general relativity to his tally is certainly incomparable.

In my opinion, the Einstein notion of inertial frames in the universe is, intellectually, the most important philosophical discovery in human history, first or second to the discovery of

---

[24] We are repeating our past mistakes with 4-D geometry.

the logic of things to create science.[25] (The logic of time--- although very important--- even if true, is part of logic and inferior to the Einstein notion of frames, which was an original philosophical insight.) The idea that any living or recent scientist could have been worthy of intellectually wiping the boots of Einstein is absurd. Only Aristotle or Bertrand Russell can be compared to him. P.A.M. Dirac thought only the Danish physicist and Nobel Prize winner, Niels Bohr, (his son also won the Nobel Prize; it ran in the family), was intellectually worthy of wiping Einstein's boots; and I think he was right.

Newton, of course, was also great but undermined his intellectual greatness with religion. Our superiors who judge intellectual greatness do not like religion very much, no matter who is concerned. What Newton attributed to God was investigated (or solved) logically by Einstein and his followers. I doubt very much that those scientists calling themselves brilliant, because they understand the meaning of black holes by simple logical deductions, could have got anywhere if Einstein's theories had not predicted the presence of black holes. But by basing their suppositions entirely on the Minkowski four-dimensional space, they are actually distorting Einstein's theory of relativity. The fact that Einstein also used the 4-D geometry in the field equations of general relativity is immaterial. I regard it as his most perfect snub of 4-D geometry and the mathematicians by writing the equation for it in his theory without believing in it---not even bothering to understand it, according to David Hilbert, a most reliable witness.

Undoubtedly Einstein must have known that, whatever might be its metaphysical status, 4-D geometry could not vitiate his concept of gravity as caused by the curvature of space. They are entirely two different concepts that do not relate in anyway. Present scientists (relying more on instruments than brains), employ 4-D geometry to determine the nature of physical reality and time. But that is serious because it makes time something quite different from what it really is.[26] The result is the view that

---

[25] Subsequently everything we do can only properly be called footnotes to Einstein. But we cannot have footnotes without knowing what they refer to and so the theory of frames must be taught in all schools. Searching for the logic of things, which we know as science, is a search for knowledge. But the Einstein theory of frames is about how life is possible in any part of the universe and therefore far more serious,

time travel is 'a scientific possibility', when, in fact, our time (a secular notion of time we get from repetitive cycles), is necessarily discrete so that it can never stretch anywhere---backward or forward.[27] Examined very closely, it seems to me that the writers of many of the books coming out about time are merely interested in suggesting that time travel is feasible; and I put the blame on religion, or man's fear of death. People are so afraid of death they just want to claw at anything, like a drowning person, that gives them hope that death will not be the end. They all hope that, even though the body may decay, the mind with all the knowledge of where the money is hidden will somehow survive. Well, don't we all? I have often wondered about the purpose of life in the universe if people are just born to die. The religions insist that it is not so. Yet, apparently, that is the situation, and there is nothing anybody can do about it. Time travel is simply not feasible because we do not even know what is time to travel by it. What we call time is nothing but how it is passing by.[28]

Talking about the fear of death, it occurs to me that death has been probed (intellectually, logically, religiously and scientifically) deeper and more rigorously than any other subject under the sun; and, being what it is, I think the seemingly endless examination of the nature of death is justified. However, there is not much left to be discovered about it.

The critical examination of death has a long history, and it has been very thorough and rigorous. Roughly sketched, man started his scrutiny of death because early in human existence it struck people that it was a pretty strange phenomenon. The main feature people could not understand or accept as natural is the fact that, very often, a person could be talking one minute

---

[26] And as space becomes dynamic under relativity, time too is said to be dynamic; yet nobody knows what it is. Everything we have used for time in all history merely shows how it is passing by only.

[27] By its mode of production our time has to be discrete. Discrete time is spent when its unit is ended. So there are no years in nature. There is only one year, repeated to be years and all the centuries. This discrete time cannot stretch anywhere; it cannot curve and it cannot travel.

[28] I must plead with the reader to be patient. I am leading on to the nature of time in the universe in this manner so that it may be clearly understood---I am preparing the secular route, as it were!

and dead the next minute and his body would start to decomposed. In his ignorance, primitive man wanted to know where his "spirit" that was there had gone to. Eventually the view grew that the soul or spirit in man survived the decay of the body, hence the notion of the "Transmigration of Souls", which taught that the air we breathed gave us our spirits, which was indestructible and went out of the body at death to return later in another body, and so forth. Pythagoras (logician, mathematician, mystic and philosopher) and his Orpheus movement were largely responsible for spreading this idea of death. Subsequently many religions borrowed some of their ideas, the most prominent of which was Heaven and Hell: eternal bliss in Heaven or perpetual torment in Hell after death, depending on how morally pure or good one's life had been on earth before death. Rene Descartes cemented this in logic with his purported logical discovery of the dual nature of man, known as "The Cartesian Dualism", still now forming part of the liturgy of many religions---even though rational philosophers have condemned it as pure fiction and senseless in scientific thought. This ubiquitous notion of 'scientific thought' (and 'pre-scientific thought' as Einstein was fond of saying) is very simple. It is conceded that science cannot explain everything; what it can do is to investigate everything rather than believing anything without investigation. After investigation what we find with the sum total of human ingenuity is what we have to live by or learn to live by. On scrutiny you will find that this simple principle brings cancer cure and relieve for many other ailments, rather than just staring at sick people to succumb by them as must have been the case before medical practice. It also brings mining and metallurgy; all manufacturing processes; roads rails and construction; printing and all the rest of what we call human civilization.

Now, Pythagoras was a pioneer in rational thought (otherwise known as logical thought), science and philosophy, but how are we faring after him all these years? Not much in rationality, I am afraid. To my mind, it is a big mistake to think that the world and people are growing wiser with all this fantastic knowledge from science and rational philosophy. Alas, we no longer live at the mercy of ignorant religious fanatics and dreamy philosophers, some of whom actually despised life and people, preaching sectarian hatred and human sacrifice. We are rather luckily inundated with thousands of newspapers and magazines, books and information leaflets most of which we discard as junk mail. There are regular television and radio talks, lectures and sermons by people who speak the language of rationality, unlike those in the past who dreamt and interpreted them as 'revelation' to be obeyed on divine authority, even though they may be as sinful as the devil. We get the academics giving us numerous ideas from psychology, biology, medicine, physics and astronomy, all trying to give us rational explanations of the world, nature, the universe and life's mysteries. We have even found the right theories to travel to the moon and back; the best evidence that our scientific knowledge is, in many respects, sound and reliable. There are some very rich people on earth with everything to live for willing to pay millions to travel outside the earth and hope to be back to enjoy their wealth; they are not afraid because they believe our science is reliable.

Yet, mainly due to the fear of death and the numerous myths associated with it, largely unscrupulous religious leaders are able to sway people more than our greatest scientists. In almost all of the major countries rulers and aspiring rulers often invoke god's blessings, or actively use divinity to win votes. Abortion, for instance, would be a practical issue without religion; having children could be rationally decided, but no. Religion is brought to bear on many of such issues even though many of the preachers know they are preaching against human welfare, namely how to guarantee the number one would be able to cater for adequately without human misery. If we add deliberate wickedness and blood-thirsty tendencies (which are pretty common) to religious stupidity and sectarian hatred, we begin to get an idea why our world is what it is despite the phenomenal increase in knowledge. So no one can claim that the world is growing wiser; in fact, when the evils of the internet are taken into account, it seems human beings are growing dumber and dumber. To use a great invention like the internet to spread ideas about suicide, mass murder, mayhem, civil disobedience, and general acts of evil is not very clever; it is a mental disease, sheer madness.

However, about death alone, I think I can give all religious people this assurance derived from science, logic and philosophy from the earliest times to the present day. It is this: you and everybody else die every day so long as we sleep.[29]

Yet we all know that sleep is sweet and deep sleep even sweeter, peaceful and restful. The phrase "Rest in Peace" was coined to celebrate this fact. Therefore death is not fearful because you won't feel a thing after death.[30] Apart from the curse of violent death, which we must all try to avoid, the actual loss of consciousness in human demise is exactly the same as deep sleep without the dreams for those who dream; you will not feel a thing after it, or even know that you are dead and no longer living. You are not going to be somewhere still conscious of what is going on behind you. That is what people are afraid of; otherwise nobody will ever know that he or she is dead, just like sleeping. So now that we do not need the religious lies, there is no reason for doing or saying anything that gives them a permanent place in our lives---they have caused enough sectarian strives in the world already.

Now, over the hundred or so years that we have had to ponder Einstein's theory, it has become obvious that additional two conditions are needed to make life on any planet or inertial frame in the universe possible---as against the perpetual whirling (spiralling or spinning)[31] of matter in interstellar space that creates cosmic gravity and eventually end in black holes as the absolute end to material reality--- such as we experience on earth---so that no life, time or the laws of physics are possible there. Thus we assume that for life to be possible we need to add time to the Einstein postulates. We must also

---

[29] Africans have more stories of dead people being discovered living elsewhere than any other people in the world, past or present. And they have concluded with the saying: "To know what death is just look at sleep".

[30] This does not mean people should rush to commit suicide in any situation. However bad life may seem to you, suicide is worse; there are always loved ones left behind who will miss you and it is not fair to them. In recent years the churches are providing social services to help the needy; much better than overfeeding and pampering paedophile priests or preaching meaningless sermons. We know now that kind and sweet words from god are no substitute for practical, financial and social assistance. Also, how the rich and lucky people live should not be glamorised on television so much. In the past envy sparked revolutions; today, I think it worsens the plight of victims of depressions, yet depressions is extremely widespread and induces victims to commit suicide, especially when they are on their own.

[31] Wherever I use one of these terms I mean all three or the concept of spinning as we think of it in Einstein's theory of the curvature of space causing gravity.

agree that not only should the laws of physics be the same for everybody, but the conditions for the evolution of life in the first place, must be there. Hence the postulates that set the metric of a habitable planet from that of a black hole may be stated as: (1) the right kind of conditions for the evolution of sentience life must be there; (2) The regular or repetitive motions and cycles that can be used to demonstrate the rate of the passage of time (without knowing what it is) must be present; (3) The speed of light must be the same for everybody; and (4) finally, the laws of physics must be the same for all.

Let me put the significance of this into the clearest possible mental picture, showing why time is also an important postulate of what makes life what it is. Suppose you are alone on an island. You are going to have to learn things or how to do things all alone to survive and prosper in the environment you find yourself in. The most essential thing is how to manage time as you realise that it is important through the day and night system. For example, you have to know by your time as to when it is safe to go into the bush, or how long to stay there and not be caught on the way by predators during the night. You will have several options. One of them is using repetitive cycles and counting the cycles as the rates of passing time---ten cycles gone and it means the egg is done and so forth. After a while you will learn to use the rates of time to plan ahead. Counting the cycles as periods you could learn that ten coming cycles will make you age ten years. This example is not different from earth time. It is not even different from atomic time. Einstein was the first and only man in history who showed how we get this stable life from the general flux of nature; and why we have to have our own time in our inertial frame.

Before Einstein's theory of inertial frames, the belief was that life is divine---but, although life does not yet appear in thousands of exoplanets, however being earth-like planets, it is likely life will spontaneously appear there, too. I can't believe that God chose earth especially for the evolution of life as a one-off event because he likes the smell of our fried bacon. Time was also regarded as absolute and generally permeating the cosmos such that a second here is the same everywhere else. But by the simple analysis of order and simultaneity Einstein showed that time is different in different places---helped, no doubt, by the Lorentz discovery of 'local time'.[32] It was all very simple: how could there be 'a local time' when all time was supposed to be generally permeating the cosmos and absolute so that a second here is the same everywhere else? It is often forgotten that this idea of general time was the rock upon which the so-called 'great religions' were based.[33] What is called "Newtonian Time" is, in fact, primitive man's time, which has now been shown to be logically untenable---as usual, Newton took commonsense as the truth without philosophical or logical considerations. He couldn't be bothered! But Einstein did try, at least, to search for the logic of anything he thought about.

No wonder Abraham Pais suggested that the theory of frames should be taught in schools. It conclusively explains in scientific terms why and how we have stable existence, *or an independent existence*, in the general flux of the cosmos. It makes mankind what he is, i.e. an independent creature able to distinguish himself from his surroundings---and therefore acquire the capacity to think, the beginning of thought about oneself and one's relation with nature. In the cosmos, or the rough and tumble of raw nature, as envisaged in general relativity, there is nowhere anywhere in the metric for stable life to evolve and flourish.[34] There is perpetual flux of matter. But once you find yourself as an independent 'Being', in the sphere of an inertial frame, it means you are in or on a body from which your independence as a separate individual with the right of existence, is guaranteed by the existence of the cosmos or the nature of things; it means you are in an inertial frame situation.

This is the best scientific definition of existence, or why we are here. It is very ingenious, extremely clever in science and philosophy; I just can't imagine how Einstein arrived at this

---

[32] Of course, the theory of inertia was known but not as it applies to planets with the two postulates.

[33] We are often told that the 'great religions' contributed to the survival of life. But they sure have had more people killed than saved. Besides, what is good in the ideas of the religions come from rational philosophers. In this sense if it is true that Jesus said anything good for mankind's survival, then he was repeating some defunct philosophers or he was himself philosophically inclined.

[34] This is one of the concepts that made Einstein a great Philosopher/Scientist.

notion. It is the ultimate of philosophic genius. The implication is that you are only a 'Being' if you inhabit an inertial frame---that is, if you are mentally aware that you are in independent existence in an inertial body. The additional concept that every inertial body has to have its time means there is no general time running through the cosmos. Given the importance of time and how it is closely associated with life, this is very important indeed as I will make clear below. Hence, in my view, the logic of time is also the logic of life.[35] The logic of time includes sentience for knowing our independence together with the ability to reckon time from the point of view of our inertial frame *only*. It does not allow us to extend our creation to other parts of the cosmos. (Of course, it is well known that cosmologists are doing this, but they are wrong, and, added to the concept of 4-D geometry, they are possibly heading for an almighty crash in the near future.) A theory of numbers is necessary, because the ability to reckon time can only be known through how time is passing and never what it is; therefore it is necessary to quantify time, and we have had the vision to quantify time gradually over the centuries as explained below, since everything has evolved.

I can see the point of view of Godel and others with similar views that time cannot exist under relativity. I can truly and honestly sympathise with them; for after all Einstein showed that cosmic time does not exist. This led Bertrand Russell to asked, if cosmic time is abandoned, then what is measured by the clock? Yet we are told by the interpreters of general relativity that time slows down near black holes. The mathematical interpretation of his theories annoyed Einstein; he was right. Bertrand Russell also addressed several questions to the interpreters of general relativity, none of which have ever been addressed. Let me state without fear of contradiction that there are two forms of existence in the universe: one form is to be found on inertial frames, and it is subject to Einstein's two postulates. The other form is assumed (only assumed) to be in black holes and Einstein did not indicate that the laws of physics such as can be studied on earth can be applied in black holes---for the simple truth is that

---

[35] For time is inseparable from life---ageing is going on all the time and so is time since it cannot be suspended. What makes time makes life. Ageing is known through time and time never stops. The two move together.

in black holes recognisable material existence comes to an end. I am not so sure that the speculations about black holes constitute desirable research in cosmology. The subject itself is not quite relevant to human existence on this plane---but astronomy is. The two are closely connected, but there is a distinct difference between them. Astronomy is dealing with materials similar to those found on earth; so it is an extremely important science. Conversely, cosmology in the sense of the mathematical 'reinterpretation' of general relativity is dealing rather with a metric where Einstein made clear that there is nowhere anywhere to sustain human life. What are we going to learn there that could enhance life here on earth?

Let us look at how the clock works in a little detail. The units of time produced by the clock are meticulously and deliberately programmed into it with consummate skill. The reason is that the units are obtained elsewhere (from the earth year), and the clock's purpose is to produce them (say the seconds usually), so that an exact number will equal to one year, or a complete orbit of the sun; it is necessary to be mathematically precise because the number of units must amount to exactly one year, since we have to start a new year or another orbit of the sun immediately; also there are all those other features of life like day and night and the moon's phases to take into account in one year. Hence the clock is not manufactured to randomly produce units of time as if by magic or metaphysically, so that critics (mostly religious) could say the clock is measuring time for our use---not at all. It is deliberately set up to produce seconds or intervals so precisely that 31,536, 000 or so would amount to exactly one year. This is important because it is a repetitive circular action. We have to do it again and again for time to continue. The ticking of the clock (any clock in existence) is not randomly measuring time for us. It is deliberately planned to do it in specific units; so it is reproducing units of time deliberately programmed into it.

In his book "The Analysis of Matter" the great Bertrand Russell (Logician, Mathematician, Philosopher and writer of genius, who no cosmologist can accuse of being ignorant of counterintuitive mathematics), addressed several questions about black holes (or general relativity) to the so-called interpreters of general relativity. As stated above, these questions are still outstanding.

Like Einstein, Russell was concerned about the interpretations of general relativity in cosmology. A subject that was, until then, the central area of the philosophical interpretation of the universe by philosophers. In those days the philosophers were using only logical inferences based on their own speculations or imagination; after Einstein, however, some scientists thought they could use scientific deductions based on the laws of physics in cosmology---and they became known as the interpreters of general relativity which is sometimes called 'Black hole', or that the black hole was its essential feature. The problem is that we know nothing of black holes other than the supposition that material existence that is subject to the laws of physics (and the other three postulates listed above) come to an absolute end there, so that there is nothing one can find out in black holes. All that the interpreters of general relativity are doing is sheer speculation in the manner of the old and discredited philosophies of ancient times. It is a disservice to call this enterprise 'science'. The laws of physics are applied in the metric of general relativity, but that is wrong simply because it is not an inertial frame answering to all four postulates; yet all four are needed for the laws of physics to be relevant in any metric—i.e. is the laws of physics the same for everybody in a black hole? And what do we mean by the term 'everybody'? Where is anybody or anything existing in a black hole? We know that even light cannot exist there.

We have to acknowledge a clear distinction between Astronomy and Cosmology. Astronomy is well-established and regarded as essential to knowing the material composition of matter on earth. The bodies dealt with in astronomy are all available for detailed study---those in black holes are not. We don't even know what is there. The best we can do is to assume that matter is continually whirling about uncontrollably in a black hole. The putative conclusion is that this matter will eventually condense and explode---but you can't build this into a definite science, and it is not correct to use physics, our definite science, there either because we do not know what is there or if anything is there at all. Suppose you discover an island from a safe distance and could not tell who or if anybody or animals lived there, would you know what to send there? On the other hand, astronomy is a field of science because the materials found in astronomical objects are the same as we have (or learn about) on earth, both practically and theoretically. Knowledge of the sun's source of energy can be safely used for power generation on earth, since similar materials are found here. Not so the black holes.

Now let us look at time. What is the nature of the time that is said to slow down near a black hole, since cosmic time does not exist? The features of our time that we all know is that it is created from point to point---the year for instance; and all other units of time are fractions of the year, as I keep reminding the reader---we must not get confused by atomic time. It is logically similar to earth-time, except that it is composed of, or occurs by, shorter pulses or units. But each pulse is a unit of time in the same way that the orbit of the sun also results in a year as a unit of time. Besides, the way and manner we create our time necessarily imply that we can only trace the rate of the passage of time and never what it is. It also makes our time discrete, year after year. We have to repeat the year because there is only one year in nature---one circle of the sun. We count the year repeatedly to get the centuries and so forth. If cosmic time is abandoned with the implication that every inertial body has to invent its own unique time applicable only to it, since the parameters employed do not apply to other bodies, which is what we call 'relativistic time', then whose time or what time is found to run slowly near black holes?

Similar questions were put by Bertrand Russell about the metric of general relativity. All we hear is the ranting of university dons, invariably praising one another for being as brilliant as Einstein because of their views about general relativity. This is very much like the old philosophies of arm-chair thinkers deciphering the cosmos in their minds by means of logical inferences---dreams, if you want to be charitable; madness, if you are inclined to be honest and damn the consequences!

The truth is that black holes are not inertial bodies or planets endowed with the four postulates listed above for normal scientific study by human beings. Nobody can have any inkling of what the nature of black holes is. You cannot apply earth time there in a proper and accurate study of the natural world without using obscure mathematics, because earth time is limited in application to the earth only, since it is based on the unique parameters of the earth.[36] Earth time has conditioned our minds to periodicities that may actually cause havoc in other metrics.

In any case, it has escaped mathematicians that, logically, using earth time (mathematically adjusted or not), to apply to other metrics---planets, moons, stars, black holes, etc.,--- amounts to proving that there is no universal time as wells as providing evidence for the fact that we create time anywhere we are or find ourselves. This should normally intensify the search for the roots of our time. Whether or not any one's suggestion is acceptable as the final word, the implications are serious, as I am trying to sketch in this book. Ultimately we have to confront the notion that there is no time---for we have to have one everywhere we go! But how is it created? We used to think that time is everywhere. Now we have several proofs that it is not necessarily so. Yet scientists continue to refer to 'The beginning of time' in reference to the cosmos. Whose time is that? The clock is cleverly structured to accord with the motions of the earth because it applies to everybody on this inertial frame. All other motions are individual activities and infinitely varied. They can only show time that has been obtained elsewhere (from the breakdown of the earth's orbit of the sun) going or moving on.

A mental defect that afflicts every scientist is that they all think there is time in nature; they speak of time as they refer to water, as if it is something they know to exist; everything they say indicates that they believe there is time. That is how they study black holes and everything else.[37] Yet, in fact, there is no longer a universal time that permeates the universe;[38] and

---

[36] You can apply earth time anywhere (and all scientists are doing it), but that is a violation of the Einstein theory of frames. Even then you can only do so with massive mathematical changes---that is what we use to land people and instruments on the moon and other planets.

[37] Otherwise why should they state that the Minkowski formula is the best way to understand relativity? So, since the Minkowski formula is false, does it means relativity is not properly understood? The theory is complex; the whole of Einstein's contribution to human knowledge is complex. Thus many aspects of relativity are correctly applied in physics; but much of it that borders philosophy, like time, is wholly misunderstood.

[38] This is not a philosophy. It is what Einstein discovered, and which Sir Arthur Eddington was referring to when he castigated Einstein's critics as 'making meaningless noises'. In other words, it has been experimentally

when we study our peculiar (or unique) time in the light of this supposition, we find that our time, like the year, being based on regular cycles, is necessarily discrete and therefore could not have had a beginning far back in the past. Discrete time cannot march. Only existence had a beginning and has been marching on ever since.

Next, like everybody else, scientists believe that they have got a logical definition for time wherever one may be. It is called "The irreversible passage of existence", by which they mean 'motion' since the whole of existence (everything) cannot be moving uniformly simultaneously. These scientists are clever and knowledgeable; most of them accept that there is no universal time but they fail to add sentience to the requirements for creating time; they think wherever one may be there would be motion of some sort, interpreted as "The irreversible passage of existence".

However, they are sadly mistaken. For however it is called (as motion or the irreversible passage of existence), there are two insurmountable logical objections for taking it as we know, call and use as 'time' in the clock. Minkowski is no help because his theory may be beautiful in mathematics but logically flawed without the possibility of redemption. The first objection is that motion, no matter how it is defined, is not one and smooth and uniform, universally moving forward (as one entity.[39]) It is multitudinous or varied and individualistic, erratic and irregular---each of us perceives the world in his own way; so if we take this route it means everybody has to have his or her own time.[40] In that case, how is the clock universally applicable in one inertial frame or planet? It is not correct to rely on intricate mathematics. People use time without having to work it out mathematically.

---

proved.

[39] That is the quality making the orbits of the sun useful as time---even then we have to remember that it is a physical motion, and, as such, can only give physical cycles. We count them as the rates of the passage of time and not the real nature of time that is why one's age is only a number of times the earth has circled the sun---as years.

[40] Maybe some sci-fi writers will be capable of inventing a society in which this is feasible, but not in ours where reliability, certainty, stability and adaptability are what make us what we are.

The second objection is also related to human perspective being infinitely varied or variable. In a world of varied perspectives, the passage of existence is not one standard thing shared by all and sundry in equal measure to give a consistent, smooth image for the clock's workings. Motion, after all, is relative.[41] You may see something move; others may see it as static; it may also be seen to move left or right by different people. I conclude that there is no substitute for quantified time, but that makes our time discrete with all the implications sketched in this book. But these cycles are physical. We count the mere physical cycles as the rates of the passage of time---the earth's physical orbits of the sun for instance. Yet all this applies to parts of the universe where human life is possible only. There is the other aspect of the universe where there is nowhere anywhere to live. Before that, let me explain the relation between motion and time. It is that, in the absence of a universal time, any time system, once created, can be applied to any motion, or applied in such a manner that any motion can be interpreted as time going as part of the natural laws of the physics of motion, but motion on its own is not time, and while certain motions can be used to reckon time (like repetitive motions), others cannot be so employed. Altogether we have to be extremely careful with the relationship between motion and time.[42] The definition given above is the reason it is possible to apply earth-time to activities (or any motion) in the cosmos whether the time is mathematically adjusted or not; otherwise there is no time in the cosmos.

You have to create the time first. So how did we (or do we) create our time---the years, for instance? To my mind, it is only

---

[41] Of course, if there is regular motions they can be used to indicate the passage of time, whereby, say, ten cycles would be needed for boiling an egg etc.---or as the time for boiling an egg; so that we would not know what the nature of time is but how it is passing only. This is the simplest way to interpret earth time, too, as I keep reminding the reader.

[42] Even ageing, regarded as chemical processing that shows natural time in action or how time moves us forward and so forth, is also subject to "The application of time", after the time has been created---that is, in the absence of universal time. Of course we have to express ageing in numbers. That implies the use of quantified time. So the means by which we express time (note it or reckon it) comes first, implying that life and time are separate entities---though closely associated. Yet it is not 'a chicken-and-egg' question. It is man who creates the time. It does not appear to exist naturally.

physical cycles we count as years or the rates of passing time. And so long as the year is seen as passing time, this theory of time seems logically flawless.

But let's put all that aside at least for now and look into general relativity and black holes where life is not possible. A black hole, obviously, is the end of material existence as far as man is concerned.[43] It is assumed that there is nowhere anywhere because there is continual acceleration and compression and spinning of matter into dense and denser states due to this constant acceleration (which we call gravity.) All matter is congealed and still being compressed so much that not even light can exist there. All this is gained through logical deduction on the basis of Einstein theory of gravity. But it means there is no possibility of gaining knowledge of any materials in black holes as we get from astronomy. By the Einstein postulates, we should expect atomic materials on earth only; that is where it is possible for sentient beings to evolve and flourish, invent time with regular cycles (not available in black holes) and have light, which also cannot exist in black holes. Light is crucial because its interactions with electrons, as Feynman showed in QED, result in the creation of material reality.

While Astronomy is brilliant science, Cosmology is not. It is sheer speculation. Any scientist reputed to be a brilliant researcher in Cosmology should be regarded with the same scepticism as used to be reserved for arm-chair philosophers. Serious scientists who study the cosmos concentrate their work in Astronomy where the materials found in the world are similar.

Here and now, I am dealing only with time such as we mechanise into clocks. A great logician like Godel could not imagine how time could be in existence and found reasons to support his view that it does not exist.[44] The truth is that Godel should have thought deeply about the physics of time in addition to its logical credentials for existence, since time is in

---

[43] The difference between special relativity and general relativity is not merely mathematical. As I have said, it shows where life is possible and where it is not. Einstein has divided the universe into two. It is a great philosophical device, or achievement.

[44] See "A World without Time..." opp. Cit. Ch.1., p6. The author, Professor Yourgrau, is convinced he is right!

daily use in all situations and activities (schools, sports, work, travel and leisure.) This is because the physics of time is clearly there and rather elementary. Physically we are using regular or repetitive cycles for time---the year, for instance. However, the repetitive cycles are passing so that they give us units of passing time only. We are gradually getting used to the idea that there are always two ways of looking at anything: the logical way and the physical way. The logical manner is the style of the human mind; the physical manner is the natural course of nature to which the human mind's logic will find strange. Thus we have particles being at two places at the same time which in human logic is very strange. Other particles have dual natures as particles and waves. Part of the reason for this dichotomy, I think, is that in the mind or psychology we deal with things as set in time. But in the physical or inanimate world there is no consideration of time since time is a psychological entity. Yet in the mind time has two aspects: the physical (say, the orbit of the sun which in nature is just a physical activity), and the internal sense of duration that assigns periods of waiting (or time through quantification) to specific units of physical activity.[45] The logical course of what the human mind expects it to be, as against the physical nature of it as perceived by the human mind or brain-eye complex. The internal assignment of duration to physical activity seems to me to be instinctive; it has been learnt over the centuries and thus has become part of our basic nature. Thus if you are told to wait for five minutes, you know instinctively how long that is going to be. But if you are told to wait for ten hours, you know you could do something in between.

I believe Kurt Godel looked at the logic of time but ignored the physics of it. Yet the physics of everything is paramount. Sometimes the two coincide and we find things easy; sometimes they don't and we find things difficult. But there is nothing we can do about it. Man is governed by his psychology, which is like a soup into which all sorts of things are cobbled together. Physical reality, on the other hand, is just how material reality tends to behave. Surely these two sources of human knowledge cannot be identical. The Physics of things

---

[45] Thus the study of time is much more complex than we think. Since Einstein showed that it is neither general nor absolute we should by now be much closer to its nature but for the Minkowski distortion.

are often unknown in the human cortex. Ignorant survival was the order of the day until science was established. That is one reason human sacrifice was so common in our primitive days.[46] Unfortunately our primitive past has imposed a sense of time on us; this sense of time makes time independent of space. At present people think that there is space; there is man; and there is time. They are wrong. Despite the new definition of space, for ordinary existence, we know of space lying about to move in. And beginning with the day and night system, men have gradually invented a sophisticated system of time and mechanised it in the clock. But what is it like in logic---that is, in the human scheme of knowledge?

For a start, we use mere physical orbits of the sun as one long unit of time and subdivide it down to the seconds. Even atomic time uses regular pulses and we count them as units of time passing---exactly like our orbits of the sun which are also always passing, so that to have more years we simply count successive orbits as 'years'. Similarly, to have more atomic time we simply count successive pulses or oscillations of the atom.

Actually, it is not true that time is always passing; the point is we cannot track it by any other means than regular cycles, which, of course, are repetitive or repeated continuously, and therefore the units that we count as the mode of the passage of time are always passing. Unfortunately that is all we can ever know of time.[47]

---

[46] Any country, especially in parts of Africa, where human sacrifice is still practiced, requires concerted efforts of succour from the civilised world. At present we all agree that the human population ought to be severely restricted to limit the number of people suffering in the world. Of course if we could make everybody happy then no limit is required. However, if we cannot do so it does not mean having children only to have them killed in primitive religions because of the difficulties of keeping them alive as is happening in parts of Africa. The so-called 'great religions' believe that they can end this vile practice, but they are wrong. The primitive religions cannot be reformed by means of occasional well-publicised short visits and gifts of second-hand clothing.

[47] From this point of view, there is something to be said for the view that time does not exist, since what we call time is created by man---but then, logically we have created something that we call time and use it for the proper regulation of human life. So there is time as a passing thing, in the sense that even if it did not exist we have created it---except that, poor souls,

We know of no other way to get time units. But it makes our time necessarily discrete (unit by unit, or year after year after year.) We recall that there is man, there is space and there is time. We all agree that the space is independent of man. What of the time? As sketched above, it is evidently also independent of man. For it is a time system deliberately created by man (as the best we can do) to apply to the external world. It is the best we can do, and it is evidently independent of its creator. To make it identical with space only by means of counterintuitive mathematics based on imaginary quantities is enough to annoy the logicians because imaginary quantities do not exist physically to be used by man.

Scientist do their intellectual standing no good at all by insisting that Minkowski's formula is the only way they can understand relativity, as some of them are claiming. Professor Sir Arthur Eddington alone (among scientists), to my knowledge, redeemed himself by describing the Minkowski formula as fictitious. It seems to me that everybody else is guilty.

In the end, we have to agree that time does not exist in the cosmos naturally.[48] It is psychological and therefore requires sentience to set the points and count the orbits or regular motions as the rates of the passage of time, which, I think, is all we can ever know of time. Yet, naturally there is chemistry, and chemical action requires waiting. We call that 'time'. Feynman, for instance, defined time as 'how long we wait', and logically he was right. Therefore there are physical actions in nature without the intervention of man that may be called time. The essential point is that down here on earth, what we call time is only the mode of the passage of it---year after year after year because we use determinate cycles to establish the rates of the passage of time, thus making our contraption necessarily discrete. So there is no need for the so-called arrow of time to move it on. The years have never needed any theory to explain how they become the centuries---they simply replicate on and on forever.

Let me stress that it does appear that time has something to do with chemical processes. Suppose you inject human sperms

---

we have just the elementary technique for showing how it is passing only.

[48] But the argument that it does not exist on earth under a relativistic metric is wrong, totally wrong.

into a woman by whatever means, and after waiting for nine months a human being is born. In temporal terms it took nine months to create the new birth. But in biology the time taken (the nine months gestation) was caused by the tedious nature of the chemical processes. However, to the ordinary man it was 'time'. Since we can never know all the chemical processes involved in any action or events, we use the word 'time' for short. The important thing is how we quantify time, or how we get it in units, as explained below.

The main reason we have had to quantify time is the day and night system, added to the fact that the year is repeated all the way to the centuries and have had to work it in such a manner that every unit of time is a fraction of the year. Thus time is known and used only in units---second, second, second; or year after year after year.

In one of the maxims with which I began this book (Postulate No 7), I said the notion that time does not run through the universe but that somebody must be there to set the points for the creation of the units of time (the years, for instance), is the really shocking new idea about time that is going to take mankind ages to understand. The reason is that everybody thinks time flows or that we live our lives moving through time which is unconsciously and impersonally passing by. Professor Anthony Flew put it succinctly in his Dictionary of Philosophy (as its editor): "Perhaps the most puzzling of the pure philosophical problem about time is that of its 'passage'. It is almost irresistible to think either in terms of its flowing or of our moving through it..." He is right. Yet since Einstein, we have realised that it is the unstoppable passage of time we need to explain, as I am trying to do in this book, because the flowing time concept is not real or true.[49] There is no longer a universal time permeating the universe or flowing about for us to live through it. Time is 'Quantified Time'. We use points to give us units of time as, in Russell's phrase, 'relation between points'--- but the process never stops, so that time seems continuous.

Unfortunately, the flowing time was the idea of ignorant, primitive man before the rise of science and logical thought. In the words of Professor Sir Arthur Eddington, "Prior to Einstein's

---

[49] Professor Antony Flew, *Pan Dictionary of Philosophy*---Ed. By Antony Flew.

researches no doubt was entertained that there existed a 'true even-flowing time' which was unique and universal...Those who still insist on the existence of a unique 'true time' generally rely on the possibility that the resources of experiment are not yet exhausted and that someday a discriminating test may be found. But the off-chance that a future generation may discover a significance in our utterances is scarcely an excuse for making meaningless noises."[50] This is strong language, and I approve of it because we are dealing with a subject so difficult that a strong language to emphasise the unfamiliar truth of the matter, even from a noble man, and from the height of the founder of Astrophysics, is necessary.

Conversely, people think existence itself is time so that as it passes time is passing. In fact, time requires points otherwise we couldn't have it in units. As I always put it, somebody must be there to set the points either for counting the atomic pulses or the orbits of the sun. Secondly, existence is multitudinous; everyone and everything is moving through nature differently—**which one is time or can be used to reckon time as the universal passage of existence**? To overcome this difficulty, we use the sun/earth system because it covers everything on earth. That is the reason even the atomic time is based on a fraction of the earth-year, namely the second, before it can make temporal sense, or give us recognisable temporal meaning.

Of course, due to the way we are forced to create our (non-existing by providential command and therefore artificial or human), time units with repetitive cycles, time, or what we know as time, can never stand still.[51] This, evidently, is exactly the behaviour of what we call time on this planet with which everybody is accustomed, namely units of time (the year and

---

[50] Opp. Citation, Ch. 1.1. This has been quoted before and it may even appear again!

[51] In defence of this chapter's title, it is argued that, given the basic nature of time (as we know it), it is bound to have a universal logic as to how it comes about or how it comes to be in nature at all, since 'there is no longer a universal time'. This philosophical condition applies to existence as well, namely, given life as we know it (eating, sleeping, thinking, falling ill dying, etc.), life in any part of the universe will be similarly affected, otherwise it will not be 'life' as we know it—e.g. can anybody outside the Sci-Fi group of writers envisage living people who neither eat nor die?

its fractions), that are always passing by because they are created continuously with repetitive cycles; yet that is all we know as time. What we know as time is what we create (or 'construct', in Russell's words), with repetitive cycles or motions. As such they are always passing by. This is a good summary of my theory of time: that is to say, the very logic of time shows that it cannot stand still because we create it, and can only create it, with successive cycles---the years, for instance.

I have to stress and repeat this several times because it is so strange, but the truth is that this process of having time raises the quandary that we can never know the true nature of time, only how it is passing by. For human beings it seems time's real nature can never be known. We base all time on the earth-year; yet it is passing. The year never stands still.

There is natural time, of course. The notion that time does not exist cannot be logically sustained---Godel or no Godel. Let me illustrate this with sports or one sporting activity with which we are all familiar: take football for instance. When the goalkeeper kicks the ball after a catch it takes time to reach the field---that is time, simple. There is no other word for it. You can interpret it in mathematics and physics, but to the man on the Clapham Omnibus it is time, pure and simple, as Einstein put it.

Again, something makes things grow; take away the chemical causes, and you are left with the period during which the growth occurs. However, that period (which is time) can only be established with points to show how time is passing by---always how it is passing only; no matter how you analyse time, that is all we human beings can discover of it. The reason is that we can only use points to establish or create periods, the same thing as time in units---the years, for instance. For the year is a period as well as a unit of time; the two descriptions are exactly equivalent. And yet the only reliable methods for achieving this is the use of repetitive cycles; which can merely show how time is passing simply because they are repetitive; they yield repetitive units or cycles. As a result, we can only know how time is passing and never what it is metaphysically.

The units of time (the seconds and years) are passing; they never stop; thus the conclusion, as we shall see, is that the passage of time occurs through the procession of time units, the years, for example; otherwise we couldn't get the centuries out of the mere repetition of one year---in other words, there are no years in nature, only the one year is repeated over and over again. This is obvious but often forgotten in the complicated interpretations of time by some thinkers. Nobody knows why time has to be discrete. The year is only one and it is subdivided down to the seconds. But why do the seconds have to tick one by one for sixty times to give us the minutes, the hours and so forth? My suspicion is that the structure of our time is influenced by the day and night digital system.

Again, the year is one unit of time (this is so important that it needs emphasis). We count the years one by one to get the centuries, or one by one to get our ages. And the year is our basic unit of time out of which all other units are obtained. This is not a theory but has been overlooked. It is in practice every day. It is true that it took astronomers and clockmaker ages to realise that the earth-year cycle is the most logical unit, together with some other features of the globe, to use for earth-time. As such, all units of time are fractions of the year, and have to be individual either by means of mathematics or the essential features of the globe---the day and night system, as I have said.[52]

However, it means sentience is required. Somebody must be there to set the points for the cycles (the yearly cycle, for instance), and further count the cycles as years or there will be no years and no seconds and all the rest of it which are derived as fractions of the year. As mere clever apes on a lump of rock, honest enough to admit that we are lucky to be here, but only able to live precariously in a hostile environment, plagued by thousands of ailments and natural disasters, unsure of anything and without help from any source, we can only track time by means of repetitive cycles due to our ignorance (using points to give manageable periodicities like the seconds, etc.), and count the cycles as the rates of time's passage through nature—it still doesn't tell us what it is. The important point is that what we know as time required sentience to 'construct', as Russell put it. He wrote, and I repeat: "It seems that the one all-embracing time is a construction, like the one all-embracing space. Physics itself has become conscious of this fact through the

---

[52] This has been explained in the essay "Time and Quantified Time" below.

discussions connected with relativity".[53] Therefore, our 'time' is not a natural phenomenon. The nature of real time is still unknown. What we know as time is a human contraption, construction or creation derived from repetitive motions as the rates of passing time---the years, for instance!

Man cannot perceive reality but only what his total equipments (biological, perceptive, physical and mathematical) can filter through to his senses for his use; real (or raw) nature could never be understood.[54] For instance, atomic and sub-atomic matter are never encountered physically; they are known through their mathematical, physical and chemical effects on man; but that does not show what they are in the absence of man and their effects on him.[55] Thus, some problems can never be resolved because how they appear to man is (or are) most certainly not how they really are in nature.[56] For mathematical discoveries are not discoveries in reality; they are mere mathematical compatibilities or what are consistent with what are known, which may be false, hence the eather debacle. Mathematics is human in origin; as such it can never discover what nature really is, only as it is filtered through to be consistent with our mathematics—which may be false---so there is a dilemma, but there is no answer. Particle physics has thrown up many unanswerable questions about the nature of physical reality. But then some philosophers have always maintained that reality is what we can perceived, not what is

---

[53] Bertrand Russell, *Mysticism & Logic*, Ch. Viii (x)

[54] There is the outside world and there are our internal tiny senses, some parody or reflections are unavoidable for the world makes no allowances for us.

[55] Professor Richard Feynman writes in his book QED: "...all there is in the world is electrons and photons..." , which, of course, is correct. I also wrote a year later without knowledge of his book that all there is in the cosmos is light energy which congeals to create all other particles and matter, because we see light plainly; it is the only thing we can see; and also we can perceive but only the light emissions of everything else in the form of their images; so the photon undermines the Platonic theory of Forms or Ideas. My book, The Metaphysical Foundation for Physics (about twenty years in gestation), was published in 1986---it fell flat.

[56] The human mind filters reality for human understanding. The world is altogether different without the human mind. How this human mind arose through evolution is the question, but arose it did. Bestowed it was not---for so far we have encountered nobody who could have bestowed it.

really there in nature, for after all we are worthless, temporary residents (mere clever orphan/apes on a lump of rock) due to disappear pretty soon. For instance (this is speculative and may even sound silly; but the quantum theory is silly anyway, so I will just state my views and keep my fingers crossed that the mathematicians will not have me lynched), in so far as the energy quotient of the quantum depends on the frequency of its wavelength, time is involved in the creation of matter through the interaction of electrons and photons as Professor Feynman is saying in his book on QED. The whole of quantum physics starts from the equation hv discovered by Max Planck and later amended by Einstein to $E=hv$. Now since frequency means 'how many times' it involves time, duration and motion. For me, the problem is this: frequency is based on our time since there is no longer a universal time, or universal time units. The second upon which wavelengths are based is a fraction of earth-time, and wavelengths determine frequency. So does the quantum exist naturally in the cosmos or merely appear so to us here on earth due to the nature of our time units upon which they are based or from which they are derived? If the latter, then what is physical reality as conceived on the theory of QED?

Thus it is not correct to assert (because there can be no logical proof of it) that the world is mathematical; that implies that man's mathematics creates the world of sense, which is not the case, because the world of mathematics is created out of the impressions received by our senses from the outside world. Most of these impressions are non-mathematical; but it is only with mathematics that man can look deeper into nature consistently as to discover reliable knowledge—therefore it is true that mathematics is the queen of the sciences, although it is not as mysterious as is usually branded about. At the end of his life, in his book, "My philosophical Development", Russell wrote an essay about Pythagoras. He said in that essay that mathematics had ceased to seem non-human to him. He called it tautological and trivial, as trivial as saying a four-footed animal is an animal, and so forth. This is true. Like all deductive processes in the human mind, mathematics has to start from something known, recognisable or indisputable, a brilliant method enchantingly demonstrated by Hercule Poirot and Sherlock Holmes (something 'elementary', Watson.) Socially the detective can be certain of his 'known knowns', but in the metaphysical study of nature nobody knows anything so certain. The scientific method, as Russell said, is necessarily tentative, but it works.

Consequently all reality (for us humankind) is like a dream. We do not know whence it comes, but we experience it and have to react to it to survive---that is reality. To the cosmos it is not important; our planet is even less than a dot, and we (psychologically fragile) humans crawling on its surface together with our period of existence are no more than a second in cosmic terms. There are so much movements and activities over such vast distances that our planet and all our creations can hardly be anything but a brief moment causing something like the flutter of a fly's wing.

Speculations as to the nature of the cosmos are futile, vain and even silly, when the resources are needed to make life comfortable for our brief existence, since man can never receive any assistance from anybody anywhere---for there is nobody anywhere.[57] Things in the cosmos happen haphazardly in a continual state of flux without time: the universe has no time. To put it another way so that no one can miss the idea, I am stressing that there is no time in the universe. There is time on earth because man is sentient; for somebody must be there to count the orbits of the sun as 'years', and use the passage of the years as the passage of time all the way to the centuries and so forth---e.g. paring the year down to the seconds, etc.

Sentience constructs the consistency and regularity[58] out of which time can be inferred, but even then only how it is

---

[57] The Encyclopaedia Britannica states (in the MACRO, please) that practically all cosmologist assume that space-time is infinite in its timelike directions. This means sentience is extended throughout the cosmos, since it is through sentience alone we 'construct' time. Yet once we use the earth-year as one unit of time that has to be repeated for time to continue (and continue all the way to the centuries), it means our time is necessarily discrete. Discrete time cannot be infinite in any direction. Besides, the Minkowski theory upon which this idea is based was itself flawed with his imaginary time coordinates.

[58] In 1997 I published my little book *Why Time is not a natural phenomenon*, which was totally ignored. Yet I think a great deal of the contradictory theories currently circulating in physics (Strings, Super Strings theories, etc.) might have been avoided if my thesis had been scrutinised carefully. I was not seeking accolades from scientists; I am just a thinker making tentative

passing, and never what it really is---because, to man, the passing years is passing time. This is the all-important proviso we should always bear in mind: if the passing years is passing time then we know only of how time is passing and never what it is. And of course we know the passing years as the passage of time, for that is how we learn of age and ageing; that is how we get the centuries; that is what some 'religious' people call 'the march of time'; and that is how we record historical events.

How can there be anything like a universal time when we are told that under relativity "There is no longer a universal time..."?[59] The irony is that scientists are dominated by mathematicians; and mathematicians generally have mystical streak. Thus, they have invented this monster called "Time Zero", a religious notion of time as I see it, since it harks back to the world of general time where the second is the same everywhere else. On the other hand, it is obvious that the second cannot be the same everywhere else because the cycles used for creating our seconds are different from those of other planets, stars and moons, etc. However, if you argue against them they accuse you of ignorance of counterintuitive mathematics and make fun of your ideas, no matter what they are because they will not bother to read them anyway. They regard themselves as the cleverest of men dealing with too important issues to listen to anybody who is not a professional mathematician.[60] Yet given the nature of time, if there was a universal time, that is, if the cosmos had its own 'generally imposed' time and our time is part of it, there would be "a programmed existence" for everything---or, to put it another way, there would be a programme for (to cover and regulate the being of) us and everything else because our time, technologies and much else---i.e. when to go to bed and when to rise--- would be the same; everything would proceed in accordance with a programme

---

suggestions for consideration. For if time is not the same as 'Being'---just being there—and also not the same thing as space, but rather is 'constructed' and therefore requires sentience, then physical reality is altogether different from what is being debated in physics at present.

[59] The reader has to bear this continually in mind that is the reason I must keep reminding him or her!

[60] Mathematicians are not good thinkers because they invent their own rules and are allowed to imagine properties when, as Eddington put it, in an experimental science we have to discover properties not imagine them.

whatever it might be, since it would involve a beginning and end scenarios, as the religious people think, and still believe in it even after Einstein. As it is, what we call time is not even time at all; rather we use any convenient cycles and call them units of time. Even tapping the finger will do. It is all very simple: if when to go to bed and when to rise are different on different planets, then universal time does not exist.

Furthermore, it is a mistake, a critical error (though unavoidable in astronomy and cosmology)[61] to apply earth time to the study of the universe---wherever you look, you must invent a suitable time for it if you can. If you cannot, the rule is not to look there at all and concentrate on minding your own business on this tiny rock we call 'home' for it would, inevitably, be so far away anyway. Earth time is based on the earth's orbit of the sun and applicable only to the earth unless we violate the Einstein theory of frames. Besides, the earth year upon which all our time is based is not even 'time'; it is merely the physical orbit of the sun we use to tell us the rate of the passage of time.

Let me stress that time requires sentience; without sentience there can be no time. Of course, logically, aspects of what can be used for time may exist. It is safe to assume that they do exist everywhere in the cosmos---like motion, physical and organic chemistry. But somebody must be there to set the points for the repetitive motions that can give the rates of the passage of time, such as the earth-year. Any system used for time would require points for the individualities of time units. As Russell has stated, the all-embracing time is a construction, a logical construction, requiring intelligence or sentience, if you like.

In all scientific literature and mathematics, time is treated as if it is synonymous with 'Being' or motion. I agree entirely. I am not trying to change normal human perceptions to something no one can think of as true of the natural world. Of course, 'Being' is time. Alternatively, any motion is time going. So being still is also time going because we age, and so forth; but chemical processes are hidden in this.[62] Besides, it is unknown, not

---

[61] Nobody has even thought of this, yet it is a serious logical anomaly that vitiates most of the conjectures in cosmology. It means the so-called philosophers of science are not up to the job of noticing such anomalies.

[62] Moreover, you know there is time going because of the clock as it keeps

available for quantification. We can imagine it is time going silently, but it is not time we can use, not what we normally refer to as time. What we can use as time must have a method, or logic, for creating periods we can expend as time for doing things---ten years, two hours, and so forth.

We do not use time as just being there; if you sit still you are not using time, and the fact that you are still ageing cannot be used as a period to plan activities or events in culture. We use time as 'time in the clock'. From "just being there" to units of time (like the years, for instance) we have had to do something to nature that could not have occurred without sentience---therefore to have time sentience is required. That something is what I am trying to describe. I call it 'Quantified Time', or the quantification of time---with the proviso that it can only show how time is passing and never what it is. It does not mean man will ever be able to manipulate fundamental nature to his advantage. What we do, or can do, to nature for the creation or construction of our time is purely mental—e.g. we note the earth's orbits of the sun and call them 'years'. However, we cannot change the orbits. Furthermore, we can never tell how long one year in terms of duration is logically. You cannot say 'so-and-so' is one year logically, without using any of the fractions of the year; but we cannot use fractions of the year to define the year itself. Often scientists speak of the year as if it is a definite period of time; yet it is utterly indefinable. The year is only a physical journey round the sun. But how long does it take to do so defined on its own as a logical unit? Nobody can tell.

What we can be certain of is that it cannot be a very long period of time since the sun is a tiny star in cosmic terms.

---

ticking on, but the clock is based on the regular motions of the earth and has been constructed artificially to show how much time is passing only. That is what I call the quantification of time without which numerical time cannot exist, and that the process of quantification makes time necessarily discrete with all the momentous implications detailed in this book. Without this artificial contraption you would not have any means of knowing that time is always passing. We mistake the passage of time for time in reality; in fact, we cannot know the reality of time. We can never deal with anything more than the passage of time on this planet, and even that is owed to our ability to quantify time, using repetitive cycles as the units of time. In practice it does not matter much, but in philosophy it is crucial. It makes time discrete.

Hence, using the year to measure the age of the entire universe seems to me logically wrong; we do not even know how long it is. Let me repeat that we cannot define the year by using any of the subdivisions of the year---such as the months---for otherwise they too will have to be logically defined, which is impossible without using the year. An original definition of time is impossible; using repetitive cycles and calling them 'time units' (like the years), is the best we can do, but they do not help to define time. Once you have life, time is the greatest mystery; for while you cannot tell why and how you have the life, at least you know it; but no one can ever know the nature of time.

Again, let me explain that we cannot use existence or motion to interpret the being of time, because, first, we need points (in both cases) to get the time as we know it (being how it is passing), yet points are our own creations---they are human in origin, which means that what they help us to create is not part of natural phenomena but artificial. Secondly, existence is also subject to time, and we cannot use existence to explain aspects of the same existence. Besides, existence and motion are experienced multitudinously---which one can we rely on as the most accurate representation of reality?

Above all, it is wrong to assert that time is the irreversible passage of existence (however the existence or the time is defined.) The reason is that if you rely on the passage of existence, you will still need the use of points to create periodicities---being the year, the second and so forth. That is the same as using the year as we do now. You will end up having a time system no different from the use of regular cycles as the rates of passing time, so that we see the passing year (the hours and seconds) as the passage of time. For, as mentioned above, the year is not even time; it bears no relation to time at all. It is a mere physical journey round the sun. The real nature of time can never be known.

We use the orbit of the sun as time just to indicate the rate of the passage of real time, the nature of which, I repeat, is unknown, simply because there is no other method we can use for keeping track of time. From the point of view of philosophy, every 'being' in the universe will face the same quandary about time. This may be called the logic of time in the universe. Every 'Being' or 'Beings' in any part of the universe can only invent a time system similar to earth time in broad outline; but they will have different periodicities (the years and seconds will be different), and, with them, peculiar differences in technologies. This is the only way to have time. Raw nature by itself exhibits no time due to the absence of (a) consistency and regularity in the perspective of a sentient being; (b) the ability to count; and (c) a theory of numbers—i.e. to count the seconds into minutes, minutes to hours, hours to days, days to years and years to the centuries, etc. For, in the absence of sentience how could the 'all-embracing time' as we have it, be constructed as Russell is saying? This is the logical grounds for assuming that there is no time in nature itself. However, on this planet we know that something causes things to grow, quite apart from chemical causes. That something we are able to quantify into time units---but only to show how much time is passing. It is utterly impossible to know what its physical nature is---my guess, for what it is worth, is that it is chemical.

Time is the most important thing in the world except life; yet even the life cannot be lived rationally without time. Thus, Bertrand Russell thought Einstein's theory of time was perhaps his most important discovery. The reader must know that, in my view, there is no 'perhaps' about the matter. It was his most important discovery, and since time is second only to life itself, Einstein has my vote as the greatest intellect of all time. My reason is that the solution to the passage and continuity of time outlined in this book is based completely on Einstein's theory of time in relativity which says "There are as many times as there are inertial bodies"—meaning that time is neither general nor absolute in the universe, so that a second here is not a second everywhere else. To create our seconds (or units of time) we have had to 'construct' the discrete time in use; and discrete time can only pass by through the procession of its constituent units. This is without doubt the most logical explanation for the passage of time without the mythical arrows of time.

I always like to emphasise at the slightest opportunity that earth time is not applicable to other worlds or metrics, especially the metric of general relativity.[63] It is possible (and easy) to use

---

[63] There is no sentience in general relativity; neither could we imagine that there are regular cycles that can be used to reckon the rate of the passage of time there. Space on its own is not time (as discussed below in one of the essays), and can never be time; we can use it merely to show how time is

mathematics to smooth out differences between the orbits and rotations of different bodies, but that is not the point. The essential point is to show the origin of time, or of our time, as proof that time does not have divine provenance; and that is that it comes from the unique orbits and rotations of the earth; other bodies will have their own parameters for their times--- e.g. the Neptune year is shorter than its day. Yet cosmologist use earth time to apply to Neptune---that is a violation of the Einstein theory of frames as the ultimate source of time, and therefore the conclusive evidence that time is completely secular in origin. Furthermore, these differences in time could have serious implications for the quantum theory and its wider interpretation in QED---it may not be the case that the quantum of energy, since it contains time elements, is universal. I suspect it could be a product of our peculiar time units.

In other works, I have dismissed the Minkowski appendage to relativity as not only useless but a distortion since it is based on imaginary time co-ordinates. I know mathematicians will want to bury me alive, yet Bertrand Russell, Professor A.N. Whitehead and Professor Sir Arthur Stanley Eddington also said the same thing. Since mathematicians are intellectually stubborn, I like to remind them of how Professor Eddington described the Minkowski formula for equating space to time[64]: "...the partitions representing his space and time reckonings are imaginary surfaces drawn in the world like the lines of latitude and longitude drawn on the earth....Such a mesh-system is of great utility and convenience in describing phenomena, and we shall continue to employ it; but we must endeavour not to lose sight of its fictitious and arbitrary nature".[65]

No philosopher will accept that space can be equated to time successfully with a fictitious and arbitrary formula and still be the method for determining the nature of physical reality, which

---

passing, but that requires sentience. For instance, even the orbit of the sun is not time---we call it a year; but raw nature has no years, just material motions or cycles.

[64] The fact that I have quoted this before is immaterial. Actually I will quote it again and again. That is how important it is, because Minkowski virtually sought to alter human conceptions of reality with a theory that is arbitrary yet scientists continue to rely on it.

[65] Professor Sir Arthur Eddington, in *The Mathematical Theory of Relativity*, Cambridge, 1930, Ch.1.1.

is what the Minkowski 4-D geometry is used for. It is a distortion of relativity which, I think, will eventually require another Einstein to weed out of physics. For all those incapable of philosophical interpretations who may wonder why I take so seriously Professor Eddington's condemnation of the Minkowski theory of 4-D geometry, or four-dimensional space, or, again, the equation of space to time, let me explain my position in simple terms. When Lorentz and Einstein discovered that time is neither general nor absolute running all through the universe so that a second here is a second everywhere else (and I must stress that it was a discovery, because it came out of experimental results), the whole of physical reality changed and mankind began to live in a completely different world never imagined to be feasible even by the great poets. Conversely, when Hermann Minkowski came in with his theory of time as being the same thing as space, mathematically known as 4-D geometry, he sought to alter the constituents of reality again. Now, if his theory or technique is assessed by our greatest experts in these matters (being Russell, Einstein, who never believed in 4-D geometry, Professor Eddington and Professor A.N. Whitehead), then we have to have the courage to reject it.

However, the sad fact is that Minkowski is emotionally supported with the whole might of the world's collective scientific establishment, with the mathematicians having the biggest say. For all scientists refer to space as 'space-time' in the Minkowski sense that space has been equated to time.[66]

In spite of the very strong condemnations from three of the greatest thinkers of the last century, including Bertrand Russell, the world's most recent great philosopher (who was also a great mathematician and could not be regarded as ignorant of the issues involved), the mathematicians claim to have three good reasons for using the Minkowski 'arbitrary' formula for equating space to time, thereby suggesting that the world is one of four-dimensional continuum. Sometimes when cosmologists are feeling pretty arrogant, they refer to the Minkowski formula as 'Minkowski Universe' or 'Minkowski Space'. Here are the three reasons as an open debate so that

---

[66] The suggestion that we may use the term meaning that time is obtained from the fabric of space is not yet generally known. At present everybody who uses the term is referring to the space-time continuum---which does not exist in physical reality, only in the mathematics of Hermann Minkowski.

readers can make up their own minds. (1) It is argued that relativity is not completely objective, because adding the time to phenomena by the 3+1 method, since the time is now known to be dynamic and variable (and that absolute time does not exist) amounts to human beings playing a part in the determination of the nature of reality by incorporating their own (probably) false time---known as "unacceptable human interference in the determination of the nature of physical reality". The reply is that it can't be helped; we are what we are and cannot assume that the universe owes us any obligation to make things easy for us; but it is the reason we have to be careful about how we define time; for that is the only logically acceptable method available to man, since the four-dimensional formula is worse (and therefore completely unacceptable) because it relies on imaginary quantities. However, once time is logically defined, the 3+1 formula will also become logically acceptable. So that the concept of reality based on that cannot be logically faulted in the way that the Minkowski formula can be dismissed for being based on imaginary quantities. (2) Next, they claim that it makes relativity easy to understand in mathematics by using one equation to represent matter, space and time. The reply is that it may seem useful, but if it is not true of the external world we cannot use it in physics of all subjects. As Eddington pointed out, if we employ it we must do so in the full knowledge that it is nevertheless arbitrary and fictitious---and what is classified as such cannot be incorporated into science particularly for the serious matter of deciding the true nature of physical reality. We might as well use soothsayers in place of experiments. (3) Finally, they say even Einstein incorporated it in the field equations of general relativity. Here we can't help but notice the paucity of brain power in physics today; for the fact that Einstein used the formula in his proposals is not a logically necessary evidence of its truth. In any case the reply is that he hated it, but was eventually persuaded to adopt it. And he did so knowing that it could not adversely affect his basic thesis that gravity is caused not by any force but by the curvature of space---whether time is part of that space or not. Furthermore, David Hilbert lamented the fact that Einstein never believed in the Minkowski formula when he said, "Every boy in the streets of Gottingen understands more about four-dimensional geometry than Einstein..."[67] I will be reminding the reader of this all through the book; not because I cannot remember that I have mentioned it before, but simply because it is crucial to my arguments against Minkowski and must be emphasized again and again.

So, then, by ignoring Minkowski, the greatest advantage of my interpretation of time is that we no longer need theories (or arrows of time) to explain the passage and continuity of time, otherwise known as perpetual time. We start with points to determine the year (or units of time since usable time is known in units only)[68], and use points, again, to show how the year is replicated together with its fractions beginning from the second, for time to pass and seem continuous, or perpetual. For the passage of the years is what we call the passage of time; but the years have never required arrows to pass by and accumulate (replicate) to become all the centuries. We use simple arithmetic. And, of course, we know that the sum total of all our units of time are derived as fractions of the year---including even the units of atomic time, or the atomic oscillations, which have always to be based on the second to make temporal sense; and the second is a fraction of the year, of course. It would not exist without it.

Discrete time means the time is known unit-by-unit---the years, for instance. Or the seconds, the hours, minutes and days. Since the day and night alternations is the origin of earth time (that is, the original source of man's sense of time), our time remains discrete. Once the year is accepted as our basic unit of time out of which all other units are derived with mathematics and astronomy, the time becomes necessarily discrete. It cannot be otherwise. The principle can even be extended to the whole universe because it seems that every time anywhere will have to be discrete. Real time is unknown. To have something to regulate life only repetitive cycles counted as the rates of

---

[67] Quoted by Professor Yourgrau in his book cited above---page six. Of course Einstein did not (would not and could not) believe in the Minkowski formula for equating space to time because, as Eddington tells us, it is fictitious, and Einstein was too clever to miss that.

[68] If you just sit there silently and say time is nevertheless going because you are ageing, when you come to tell how much you have aged you will need time in units; so how we get the units of time is the important thing about time overall---alas, we start either with the seconds or the years (at both ends), in units, and in units only.

passing time are available---so that is time, 'pure and simple'. To my mind, that is part of the logic of time: we do not and cannot know what time is. To get any idea of what it might be to live by, we have to use repetitive cycles and count them as 'time units' passing by. What amused me when I began thinking about time is the fact that academics are wasting resources theorising about how time passes by when the years have been passing by every year through replication. And we know the passing years as the passage of time!

Yet the logic of time, as discussed in this chapter, could not have been conceived without the concept of 'inertial frames' as the habitat for sentient beings who are able to flourish and found civilizations. Therefore, in conclusion, I like to emphasise the unique philosophic importance of Albert Einstein in addition to his incomparable science. To illustrate what I mean, I quote from Professor Bernstein's book, "Albert Einstein---and the Frontier of Physics", previously cited in all my books on time (at least for those readers familiar with my work.) Professor Bernstein writes, "In the absence of gravity, space and time are distinct entities. In the metric of special relativity they play distinctive roles. But in the presence of gravity the metric is altered, and space and time become mixed up with one another..." The unique philosophic genius (certainly uncanny and unprecedented) is the metaphysical distinction between the general relativity metric (or interstellar space) and the special relativity metric (or perspectives in an inertial frame as on a planet, where gravity is weak enough to make time distinct.)

Generally, we are told that special relativity is a special case of 'relativity', as if relativity is basic to the universe and our inertial frame is a portion of it. Perhaps the term 'relativity' is even unfortunate. As it works out in practice, philosophically, 'Special Relativity' refers to a special, calm place for life to evolve and flourish, or for living in the universe as a habitat (our home here on earth, for instance), that is different from that of general relativity which refers to general motion and perpetual flux in nature causing gravity so strong that it is not a place for living or any sustainable, sentient life. This is the correct interpretation; but relativity is generally misunderstood.

Mathematicians interpret it differently; so do the philosophers; and the physicists also, who claim that it is their speciality, arrogate the right to interpret it for everybody else. Yet, because of their adoption of the Minkowski fiction as the best way to understand relativity, it seems physicists have also got it as wrong as the mathematicians who never awoke from their slumber over the beauty of the Minkowski mathematics! To tell them it is false is to risk the possibility of being lynched if no one is looking. That is the reason I have repeated the judgement of Professor Sir Arthur Eddington, the founder of astrophysics, that the theory is arbitrary and fictitious.[69] I do so deliberately to hide behind his illustrious image and escape the inevitable harsh condemnation from mathematicians.

Unfortunately, Professor Eddington deemed it necessary to insist that mathematicians shall continue to employ the theory, with the caveat that they should, nevertheless, remember that it is arbitrary and fictitious. When he died, everybody forgot about the caveat that is the reason Minkowski still reigns in the determination of physical reality with his totally fictitious formula for merging space to time. And I put the blame at the door of Professor Eddington. Why should he urge scientists to use a theory he described as arbitrary? How can a fictitious theory be used in physics for the serious matter of determining the nature of physical reality?

When Professor Sir Arthur Eddington (and those of us living in Britain know how status matters) has decided that a theory should be used, who is a black man--- from the jungles of the old British Gold Coast Colony in rain forest Africa--- to tell them not to do so because the theory is false? What about the logic?

---

[69] Let me remind the mathematicians that the only benefit they think the Minkowski formula confers is, in fact, illusory and dangerous. It makes time seem to be already implied in space, so that s=ct can be used in all equations and theories and it will hold good implying time in any way desired. This is an attempt to cheat nature and logic. The fact is that the formula is based on imaginary quantities and therefore it is not true of the world. It holds good because time cannot be suspended; but to know the time you've still got to consult the clock which operates on the principle that time is independent of space, therefore s is not equal to time (s=ct is completely false). The contrary opinion makes reality seem quite different from what it really is and that cannot be used in physics---yet they do; all mathematicians and cosmologists instinctively do accept, and theorise on the basis that s=ct; and for that reason I believe they are distorting relativity and physics as a whole. In general relativity it leads to quite bizarre suppositions from cosmologists.

Samuel K. K. Blankson

What logic? In Britain, status matters in all cases---even in prison.

If I were an eccentric nobleman, ah, then, perhaps somebody would condescend to spend a few minutes to look at the logic of the matter to make a name for himself, but not otherwise. Also, there are bread-and-butter matters to consider. I mean the money and fame aspects of intellectual life, scholarship or scientific research that sometimes resemble cloak and dagger warfare with unholy ferocity and crass ruthlessness: the research student who discovered Pulsars did not receive the Nobel Prize for it; her supervisor did. I am here referring 'only' to all those tomes praising Minkowski that need sales, including some from Einstein himself.[70] David Hilbert has confirmed that he did not believe in the Minkowski formula as logically valid. But in public he praised it. What game was he playing?[71] It is not right that matters of physical reality are treated as some sort of intellectual joke. Philosophers, as difficult and incomprehensible as they often appear to be, nevertheless display serious attitudes to matters of reality; they will never treat them as some kind of mental joke as scientists are treating the Minkowski attempt to alter the basis of physical reality from the 3+1 formula to one of four-dimensional space, or 4-D geometry---even though it is false, and we all know it. At least Professor Eddington and Bertrand Russell told us so.

Personally, I suspect the Minkowski theory is popular because it enables mathematicians to speculate that time travel is possible, being thinkers with celebrated mystical streak. In calling my theory of time the logic of time in the universe, I know very well that it will attract criticism even before it is examined; and so far nobody wants to examine it at all.

---

[70] I am aware that aspects of my theory scattered in ten books (especially about the fact that man's tiny brains cannot decipher the secrets of the universe) have been plagiarised by some very high-up scientists who are so big that I dare not mention their names.

[71] Perhaps Einstein was one of the most skilful political manipulators that ever lived on this planet. Well, I think so; he knew how to silence his critics with consummate political skills. Replacing "the usual time co-ordinate t by an imaginary magnitude $\sqrt{-1}.ct...$" is surely inadmissible in physics, yet he wrote it to gain the support of the mathematicians who were criticising him. We are now told that he actually never accepted it. That is consummate political skills for me.

Nevertheless, I insist that time cannot exist in general relativity. It can exist only on an inertial frame; and it can be created only by sentient beings because, "There is no longer a universal time..."

This means we have to find out how our own time began; how it continues, and how it passes by. In addition, we know that the earth will cease to be capable of sustaining living organisms like ourselves when the sun dies. So, what will happen to our time? If our time will die with the earth then the nature of life is altogether different from what we think it is. The first casualty as the intellectual negation of its salvation (of the salvation of myths), is religion. We like to think it helped mankind to survive; in fact, by the time organised religion was able to pretend to be making meaningful contributions to human welfare, there were (and had been) great philosophers and other serious thinkers. It is the ideas of the thinkers and philosophers that helped us to survive the darkest ages in history to end things like human sacrifice; after all, the religions also brought sectarian conflicts still going on today in some backward regions and tribes---e.g. 'Love Thy Neighbour as Thyself' and the Saints (including the influential Paul and Peter) came late, long after Socrates, Plato, Aristotle and countless other thinkers.[72] In fact the saints were not very good thinkers since they merely preached religious myths bearing no relations to observed facts---the son of God coming down on earth to save mankind indeed. This stupid idea is responsible for people looking up to the sky, as if to Heaven, in delight or supplications.

At his second inauguration, that lecherous man, Bill Clinton, kept looking up to the Heavens; cynics would say that is why God punished him. Again, the Christians now like to portray themselves as tolerant and enlightened. They appear so because the philosophers have helped them enormously to perfect their ideas of humanity; and also because other religions are more backward, intolerant and cruel now; but so were the Christians in the past upon the basis of a doctrine that is patently bogus, fraudulent, foolish and sickeningly immoral, including the burning of witches (still carried on in some

---

[72] See the relevant sections in Bertrand Russell's *History of Western Philosophy*.

Christian churches in Africa and elsewhere),[73] the persecution of scholars and so forth. No religion can be adorable because none can find a logical basis as a motivational force. The worst trickery is the mere dreams called 'revelations' by which deranged people impose their foolish ideas on the rest of us as gospel.

However, this digression has gone on too long. As the reader may have noticed, this is not a coherent technical book; I have been ignored for more than fifty years so I am not hoping to gain any honours with good behaviour. I planned this book as some sort of 'mopping up' exercise of my ideas about time so that, at 74, if my life is lost through the many ailments afflicting me, interested scholars may have a good idea of what I have been raving about.

In general relativity, there is no time because there can be no stable perspectives for spatial regularities that can be used to track time. We are lucky to have the day and night system together with regular orbits of the sun. Thus, existence (meaning conscious beings) cannot have any of the parameters necessary for survival in the metric of general relativity. It means life can only exist and flourish "in" or "on" an inertial (special relativity) frame. This is the ultimate of metaphysical genius---to discover the conditions under which life is possible in the universe. It defines existence, sentience, consciousness and time (since time requires regularity, points and sentience), without which life cannot flourish. There will always be some kind of embryonic life even in general relativity, but not such as can flourish beyond a certain primitive level of existence. To discover this as a condition of existence throughout the cosmos by means of provable scientific thought is certainly the ultimate of genius---in science, logical thought

---

[73] Of course, the witches are not burnt at stake in these modern times; the philosophers' liberal laws prevent that; but innocent people are still eliminated by many other subtle methods, including poisoning with drugs, especially children, the infirm and old people. There is evidence that some churches are committing such murders even in Europe and Britain. One woman has reported that because she failed to close her eyes during prayers, her forehead was rubbed so vigorously by the pastor till she collapsed and fell into coma and nearly died. Presumably it is 'an act of God'. She recovered six months later, but nobody believed her story and the criminal went scot-free.

and philosophy. Yet this is only part of the reason we regard Einstein as "Philosopher-Scientist".

What I find most interesting is that the Einstein theory of frames (bigger, more useful, profoundly philosophical and infinitely credible than the Platonic Theory of Ideas), was actually discovered out of the strange behaviours of time, showing altogether how fundamental time is in human life. When Lorentz found that time is different in different places (known initially as "The dilation of time as a measure of moving clocks"), Einstein single-handedly worked it out in his noble mind to mean that different frames have different characteristics and parameters, and so forth. Thus I am happy to add my little voice to the call from Abraham Pais that the theory of frames deserves to be taught in all schools as one of the means for the promotion of the scientific thought process. It is not the science that matters so much as the scientific attitude to examine every aspect of reality rationally. God knows that the applications are often left to politicians to mess it up, but that is another matter.

Because of the theory of frames, we now know that there will be life in other parts of the cosmos because there are billions of planets like the earth. Such lives will have time; and the time would be roughly like the one we use on this planet, but uniquely created with their own parameters that may not necessarily (in fact, they cannot) be like our own unique parameters. My suggestion is that the logic will always be one of using repetitive cycles and counting them (like the years) as units of time, pared down to convenient units like the seconds and so forth---that is the logic of time in the universe. This is credible; the alternative is to assume that time was bestowed on us by a Deity the provenance of whom is unknown, and still insist on a form of life (humiliating, even dangerous to us, like human sacrifice) as homage to the Deity---Einstein didn't like that---because our own fellow men are imposing these conditions on us from their own minds, and for their own, certainly, deranged or selfish ends. However, I have said enough about people and politics; but I take the view that time is of the utmost importance to people and their politics because it involves everything we human beings do; and I also want to encourage good people like Bill Gates and discourage the paedophile priests who deceive us with their meaningless dog collars.

2

THE NATURE OF TIME BEFORE AND AFTER EINSTEIN

Before Einstein time was supposed to be general and absolute such that any unit of time here is the same everywhere else. After Einstein it is not so. Rather we see it as limited to a frame and discrete because we can reckon its passing only with repetitive cycles. As many writers have observed, Einstein's theory of time arose from experimental results and therefore not open to doubt.

TIME BEFORE EINSTEIN

Generally speaking, few writers have studied time seriously with a view to suggesting ideas about its nature before Einstein. Even those who made such attempts, like Henry Bergson, were guilty of assuming that it is just there.[74] That we

---

[74] Henri Bergson, a great French philosopher in his day, was one of the very few brave thinkers to write a book about time, though he spoilt it with thoughts about 'Freewill'---the two don't mix very well. However he regarded space and time as completely independent of each other. They still are.

find it in existence, and that is that. Even the careful logical thinkers fared no better. They made it look synonymous with motion or 'Being', eventually calling it the "irreversible passage of existence". Yet existence is not one; it is multitudinous and individual. And every individual is uniquely separate with his or her own perspectives---no two persons, as Einstein showed with his analysis of simultaneity, perceive one event identically---space and time coordinates are involved. The multitudes of people perceive the world differently. Above all, existence is not altogether passing in tandem. In quantum physics directions are not even fixed; what may be irreversible to you could be the opposite to somebody else looking at the save event from another angle. Thus time had many different meanings for different thinkers before Einstein. I identify four such meanings all of which, due to the importance of time, have become the focus of mass consensus, following or even civilization. First, we have the brilliant and profound, world-shaking, intellectual interpretation of time proclaimed by the Irish Prelate, James Ussher that God created the universe at exactly noon on AD 4004; it set time in motion throughout the cosmos which has been marching on to give us the story of history. Let us call this religious view 'Act of Creation'. Even school boys can see that it has not much intelligence to recommend it. Yet the majority of mankind, numbering more than seven billion seriously believe that this is how the world came to be in existence, and worship anything they believe to have 'Created' it.

Next, Isaac Newton. Newton believed in absolute space and absolute time, and since he was very great in science, everybody able to think followed the great man's definition of time, and it ruled scientific thought until Einstein. The third movement is the Lorentz/Einstein denial of absolute time and absolute space, demonstrated (or proved) with scientific experiments. Soon after that we got the fourth interpretation of time, called the Minkowski formula for space-time continuum, or 4-D geometry for short. Its brevity belies its momentous effects on scientists. Because it was supposed to be scientific, or the mathematical interpretation of the Lorentz/Einstein experiments, practically all scientists refer to every space as 'space-time', meaning that space has been equated to time---that the two entities have become one. And yet he could not even define time. Also the very great scientist and mathematician who confirmed the general theory of relativity, as we have noted, Professor Sir Arthur Eddington, the founder of astrophysics, described it as arbitrary and fictitious. I have therefore rejected the Minkowski formula as logically untenable, since it is based on imaginary time coordinates. The 4004 Creation of the universe is too dumb to think about, and Newtonian absolute time is abolished by Einstein, so we are left with the rational consideration of time Einstein proposed, namely that time is derived from your local space---points and sentience are required, exactly the way we obtain the year, which is the basic time unit on this planet. For the philosopher, the problem is not that time does not exist since we are using it daily in all sorts of activities, but how we get it, how we 'construct' our time, as Russell put it, because cosmic time is abolished by Einstein. This is the current logical situation about time overall---that is, scientific, philosophical and practical. Now let us look at the specifics, or how some writers have considered the matter.

With regard to the view that time does not exist at all, my understanding of the thesis of Professor Yourgrau's book mentioned above is that the legacy of Godel and Einstein to the effect that time cannot exist under relativity has been scandalously neglected or forgotten by the world. Additionally he says, when that is given its proper due, time travel becomes a scientific possibility.[75] But this cannot be true because the very idea that there can be such a thing as 'The Logic of Time', as proposed here, is derived from Einstein's researches; this logic makes time necessarily discrete, so that we count individual years to get the centuries and our own numerical ages. The fractions of the year too (the seconds, hours, minutes and days) are also separate and individual units of time. Obviously, there is no chain in time; every unit is uniquely separate. That is what is meant by discrete time. However,

---

[75] It is difficult to see how time travel could be possible if time does not exist. It is also not an honourable intellectual stance to argue that something human beings cannot even avoid does not exist. It may be intellectually difficult to define but that we know something we call 'time' is indisputable. But then we clearly have to understand that intellectuals have a bounding duty to explain the reasons why and how an indefinable entity is in daily use. That is the situation with time, God and life. Of these three basic conundrums time alone seems (not only in existence) but positively decipherable.

discrete time, such as we have here on earth, makes suggestions of time travel something like mythologies. Yet the sad fact is that most of recent writers on time seem to be only interested in theories that make time travel appear to be, as Professor Yourgrau put it, "a scientific possibility". Or to quote him in full: "Godel, the union of Einstein and Kafka, had for the first time in human history proved, from the equations of relativity, that time travel was not a philosopher's fantasy but a scientific possibility."

I want to assure the reader that there are absolutely no such (logically valid) equations in relativity. In special relativity, there is none at all, and yet that is what concerns us most, since there is nowhere anywhere in general relativity to be capable of giving time there to anybody. The so-called interpreters of general relativity carry earth time there in clear breach of the Einstein theory of frames. What happened is that mathematicians coerced Einstein to incorporate the Minkowski formula for 4-D geometry into the field equations of general relativity---yet the Minkowski formula is logically flawed and therefore completely unacceptable as the basis for altering physical reality to one of four-dimensional space, in fact, it bears no relations at all to physical reality and looks beautiful only in his own mathematics, which is considered arbitrary.

What the mathematicians wanted was a new formula to replace the 3+1 system so that, as they do now (even though they know that it is not logically valid), they could write one equation to represent time, space and matter as '$s=ct...$' It was an extraordinary demand to address to inanimate nature, and would be funny, something like a silly prank, if it were not so serious. Man is an orphan because he does not know whence he came. He is a clever orphan/ape on a lump of rock; and for such a creature to demand something from nature by way of his own mathematics is nonsense. Yet they went ahead, and Minkowski obliged. However, Professor Yourgrau himself has quoted David Hilbert as saying Einstein did not believe in the concept of 4-D geometry or four-dimensional space. Therefore, space and time remain separate entities, especially on any inertial frame like the earth, as Einstein made them in special relativity.

In all history, time has been regarded as the same as existence, hence the age-old definition of time as 'the irreversible passage of existence'. In effect, it equates motion with time. In addition, since nobody knew (or still have any idea) how life came to be, no one worried about the nature of time which is closely associated with it. Instead time and life were lumped together as constituting the eternal mystery on earth.

The mathematicians used astronomy (being the natural features of the world, as perceived) to construct clocks, and that was it. We simply used the time provided by the clock makers. This time was taken as general and absolute; it was assumed to have originated from divine sources; and since a centrally imposed time system could not be different in different places, one second here was supposed to be one second everywhere else.[76] Then Einstein burst on the scene with his new idea of time.

Given the supreme importance of time in human affairs, the irrational view of time (before Einstein, as roughly sketched above), influenced all life and all ideas including religion and historical narratives adversely, spawning mythologies many of which are still with us today with their own consequences, some of which are even detrimental to human existence. One of these is that history is the march of time---marching since the 'Dawn of Time'---instead of recognising that history has been the story of how life has been lived through successive events since the 'Dawn of Existence', or intelligence, or the first acts of sentient beings on earth, from which acts all successive events have flowed as the inevitable consequences thus giving us the continuing story of human history on this planet. Now, rationally, time is not seen as marching on and taking us with it; rather we think we are marching on event by event and recording them as they occur at certain dates and times.

TIME AFTER EINSTEIN

Albert Einstein changed the debate about time for good with his division of the universe into two distinct categories, governed by different natural laws: (1) General existence, or general

---

[76] Nowadays, I guess no educated person will believe that one second here is the same everywhere. But in the past even Newton accepted the religious view that time is absolute or fixed for the whole universe.

relativity, where objects or matter just existed and whirled around under the influence of gravity without any conscious directions; and (2) special existence 'in' special relativity frames or bodies, where the two postulates and time applied, perspectives arise and intelligence and life can flourish in response to the intelligent use of available resources for civilizations to rise and fall---or generally for life to flourish as it cannot do in the whirling flux of general relativity; thus creating the never-ending chain of events known as history or the continuing story of human existence. Since civilizations arise upon definitions of time as mentioned above, the reader can see that only the very rational, scientific civilization can be consistent with this new concept of time.

Einstein did not deliberately set out to change our view of time.[77] It was an accident discovered by Lorentz. He said the Lorentz concept of local time may be regarded as 'time, pure and simple'. His genius made it sound simple, but it was the beginning of the most profound revolution in human thought. It was unique; the nearest idea pointing to the origin and purpose of life because time is the second most important thing in the universe, bar the life itself.[78]

Of course, on the other hand, Einstein did deliberately (and even contrary to classical physics), set out to change our views of the universe. The result was his theory of frames, as mentioned above. It divided the universe into two distinct categories. One is general relativity, where there is nowhere anywhere for life to evolve and flourish; the other is the inertial frame, where life is possible and civilizations can rise and fall.

---

[77] Time is the essence of human life and the cause of the rise of all civilizations, but the irony is that, what we call time is only how it is passing by! To define time to the bottom is to find the essence of life—which is physio/chemistry, motion and intelligence all taken together. There is no doubt that the familiar ingredients of evolution were present. The mystery is how they came together so perfectly to result in a sentient being with intelligence to probe nature and improve his life with the materials found on the planet without any help from any quarters. Of course religious leaders claim to have the answer; but it turns out to be just what they imagine in their dreams---the reason it is considered foolish for gullible people to worship any god at all.

[78] If time is physio/chemistry then so is life, as the theory of evolution makes clear because the two are very closely related and inter-connected.

On the nature and passage of time and 4-D geometry

Time is required in this second division of the universe; and the local time idea was just the thing to suit inertial frames. I think we should now write time as the third postulate to the two postulates of special relativity.

The problem thence is to discover how our own time began, not as a version of a universal time. That old idea was a mistake; yet everybody in science is still considering time as if it is something generally in existence and our time is a version of it.[79] Thus, the Minkowski formula for 4-D geometry is defined as incorporating time in the three dimensions of space to create 'space-time', the merging of space with time the end result of which is to give us what we call time as 'space-time'! In the absence of a universal time, where is the time incorporated into the natural dimensions of space coming from, if the end result is only to create time again? Using time that does not exist naturally or universally to create space-time as time by means of mathematics--what sort of logical reasoning is that?[80]

In fact, as Russell put it, "There is no longer a universal time..." Thus, he asked, "What is measured by a clock?" Yet the question is wrong. The clock does not measure time. It rather reproduces units of time specifically programmed into it for reproduction. That is the reason it works in units only---second, second, second, and so forth. The real problem is how the units of time programmed into the clock are derived in the absence of a universal time. This was answered by Russell himself though he did not realise it. He defined time and space-time as 'relations between points'. He was right. That is how the seconds and years are created.

The year, of course, is basic. The seconds and all other units of time are derived from the year with mathematics as fractions thereof. Everything depends on the use of points. We use points to get the year. The fact that it is repeated over and over again to give us all the centuries means our time is determinate---in other words, our time is discrete. The essence of a discrete time is ended when the units are expended, thus

---

[79] Culturally it does not seem likely that we can ever abandon all traditional references to time; but in analysis scholars should try to do better.

[80] Yet this space-time is compounded of the materials found on the earth for the convenience of the mathematician, according to Bertrand Russell, who, of course, did know a few things about mathematics.

we have to repeat the yearly cycle for our time to continue all the way to the centuries by replication. In addition, the system runs all through our units of time: from seconds to the minute, minutes to hours, hours to days and so forth. Our time is not a thread running through nature as we used to think; what we have found through experiments is that it consists of a chain of individual units created with points or mathematics in association with astronomy and the essential features of the globe; as such it consists of separate moments, as Professor A.N. Whitehead has stated in his book, "The Principle of Relativity", which means we only realised this after relativity.

It also means time is not known ahead; what we call time, say the year, is known after it has passed---e.g. the year is not ended until 31st December. That is when we can have a whole year. Then we have to start another year. The same principle applies to all the other units of time derived from the year, including, as I keep reminding the reader, the atomic units of time, because they have always to be related to the second to make sense. Secondly, time cannot be seen as the cause of events; events are physically caused; the times are recorded as the period during which they occurred. Thirdly, time created with points and which is not part of a universal time, cannot have anything even remotely to do with what the religious leaders dream up about the nature of time. Fourthly, the passage of a time system produced with points unit by unit, as the year shows, requires no arrow or arrows to pass through nature: the units replicate to pass by---precisely as the years replicate to become centuries. All that remains for time to take its rightful place in science as a rational subject is for mankind to wean itself from the 'sweet' religious suggestions about time (what Professor Eddington called 'even-flowing time'), since the true facts are now well known: we don't know what it is, except to guess that it is the product of sentience, physio/chemistry and motion; but we know how it begins; we know how it passes by---second, second, second; or year after year after year; and we know how it will end, that is, when our planet ceases to support sentient beings who can count the orbits of the sun as 'years'. Religion has nothing to do with it. The arrows of time for its passage through nature is redundant; and it is definitely not universally existing in the cosmos because without knowing how to count the orbits of the sun as years, there could be no years just bland existence.

In conclusion, let me point out that, following from the postulates in the Introduction, if the distortion from the Minkowski formula is eradicated, the question of time under relativity becomes simple, exactly as Einstein put it, namely 'pure and simple'. Here are the basic facts: (1) There is no longer a universal time so we have to search for the origins of our time because; (2) every 'body' or inertial frame has got to have its own time ; (3) under relativity the all embracing time is a construction, like the all-embracing space; (4) both the earth-year's time and the atomic time use regular or repetitive motions to track time----that means they can only track passing time since the pulses or motions can be counted 'after' they have occurred and not before. We put all this together and get the notion that time cannot be logically defined, which means that what we call time are units of passing time. They are units because we get them from repetitive motions or cycles---and that is the reason it is passing time, simply because, of course, these regular cycles are passing. One after another (or year after another year), there is nothing more to time. All units of time are derived from the year; even the atomic units are related to the second to make any temporal sense.

Thus, in the end, since the years are our only means of noting the passage of time, the explanation of time was rather easier than going through all those complicated mathematical and physical theories of arrows, mysticism and divinity. It is conceded that time is mysterious. It is even assumed to be the last refuge of God, since many people believe that time's obviously deep and fearful and intimidating mystery goes beyond human comprehension.[81] Yet it is rather ironic that we have been using the orbits of the sun for time without realising that it is all of our time---mere physical cycles counted as years, centuries, millennia---because we thought we measure our version of time out of general time permeating the cosmos, the provenance of which was assumed to be nothing but divine.

---

[81] Time is the closest thing to the nature of life. Whoever solves the mystery of time must know more about life than the rest of us, and that honour goes to Albert Einstein for observing that the Lorentz 'local time' notion can be defined as time, 'pure and simple'. Until then everybody assumed that time originated from the cosmos and probably of divine provenance. If anybody can invent time, or his time, then its origins must be human. And that was a great philosophic insight.

Yet once we learn that there is no longer a universal time (thanks to Einstein), and that we do not measure time at all, the orbits of the sun appear in a new light, namely we count the mere physical orbits as our ultimate units of time (the years), out of which all other units are derived..

3

PHILOSOPHER/SCIENTISTS

Plato allowed for a creator. His theory of Ideas is one justification for a creator. But he was wrong. A Theory does not create reality; it can only reflect it through physical evidence. In the absence of such evidence, all theories are matters of opinion. Every suggestion from a human being without physical proof should never be accepted as worthy of attention. Plato has had adulation from religious people for far too long. Even Einstein never got a fraction of that kind of intellectual adulation; yet he rather had the necessary physical proofs. Human beings have been governed by some people's mad dreams for too long---chief among whom is Plato. In Hellenic times, as opposed to Plato, Lucretius rather was right. The quantum theory has proved him right: particles of matter are continually on the move even within a single atom. As they do so (or 'swerve') they accidentally cause chemistry; one result is life. Those academic philosophers still writing footnotes to Plato should be ashamed of themselves as Sir Karl Popper has observed.[82] In what follows, I have quoted passages from

---

[82] What Sir Karl Popper is moaning about is nothing but academic philosophy, the very subject he taught his students throughout his illustrious career. What I think he is hoping for, due to the influence of Russell and

scientific writers and others from philosophers and have contrasted them to show how some philosophers (who are not followers of Bertrand Russell) continue to indulge in the fun of bypassing scientific activity as if it does not exist.

In the past philosopher/scientists were born not made, but since relativity, the quantum theory and the electronic revolution, the internet, the computer and all the rest of it, every aspects of life has become so complex that we need to train our own philosopher/scientists to take care of us, especially in medicine. Socially too things are getting out of hand, so we have to be careful or life can be extinguished very easily through the actions of a few mad men. The artificial or classroom philosopher/scientists we can create may not be as brilliant as the originals, certainly not, but equally we cannot go on like this without their guidance---for instance, relativity is still no properly understood. Mathematicians claim that it is only with the Minkowski formula they could make sense of it all; on the other hand logicians insists that since the Minkowski formula relies on 'i' or imaginary time, it cannot be used to determine the nature of physical reality as to whether or not the physical world or physical reality is four-dimensional. So what is happening now is a scientific anomaly because all science is based on the concept of four-dimensional space, as space-time, yet logically the notion is fatally flawed. I count myself among the very few people who are writing against it amid mockery from The Royal Society's administrative minions.

Already we are able to train scientists and philosophers; even self-educated people have to learn their ideas from books; that is also training of sorts, except that self-educated people have to work harder. All educationists know that what a professor can easily impart to his students in one lecture would take the self-educated person ages to gather from books. Nevertheless, we train all our experts (physicists, astronomers, mathematicians, biologists, bio-chemists, architects, engineers and dozens of others in technically demanding professions.)

Einstein, is an end to the intellectual habit of bypassing science. But if we are henceforth to give due respect to science, as Russell said, "on pain of death", then everything is bound to be footnotes to Einstein, particularly because of the theory of quantum which is also (in theory) due to his paper *"On a Heuristic Point of View about the Creation and Conversion of Light."* Under QED light is the new reality.

They don't come from above; training makes them what they are or what they turn out to be. Many of them go on to teach others in schools, colleges and universities. It is true we don't know how to train people to make discoveries and inventions or propound profound theories about the universe and the world we live in. In a word, we do not know how to create geniuses. Still we do from time to time get some people to whom discovering something out of the usual or from thin air involves sacrifices they would readily subject themselves and their families to without thinking of the money, fame or even the preservation of their lives. From such persons we get our ideas, useful ideas, I must stress, and so I will quote three passages about scientific facts and other three about philosophical ideas imparting great wisdom to illustrate what knowledge means to us. This is a book about time based on relativity, so I think it is right to demonstrate the merits of the scientific outlook as against the Platonic mysticism.

First, a scientific piece (not a mythological one in ancient man's style) about the sun itself as the source of life on earth: "Most of the energy of the sun comes to us in the form of light. Sunlight is transmission of energy, in the form of electromagnetic radiations from the sun to the earth. When Stephenson's first crude steam locomotive was moving along its wooden track, the inventor asked one of his companions what was driving it. 'Your engineer from Newcastle, I would say.' 'Wrong', replied Stephenson, 'the sun is driving it.' I suspect that many millions of us who race the roads in our automobiles have not yet grasped the meaning of this simple sentence. Most of the mechanical work of the modern industrial world is done by energy stored in fossil fuels. Other power comes from water lifted aloft as it flows downwards to the sea...Visible light is the most familiar form of the radiant energy that reaches us from the vast and distant sphere whose surface temperature is estimated to be about 10,000F..."[83] These statements are all facts, even though they may sound strange to the layman. But how have they been pinned together? The answer is bit by bit over several centuries and by numerous contributors. The important thing is that they are true or cannot be refuted without further research. The research will have to be scientific; you

---

[83] Professor Paul B Sears, *The Biology of the Living Landscape*, Allen & Unwin, 1962.

could not do it sitting comfortably in your armchair. But equally you do not need to go into a laboratory to do the necessary research. It is a question of attitude. Your thinking must be scientific as stated by Bertrand Russell in his essay "The Rise of Science", namely: "It is not what the man of science believes that distinguishes him, but how and why he believes it..."[84] How he believes it is more important---how he believes what he does, what actions did he take to lead him to those concepts? It is reported in the Press that many people were surprised by the success of the recent science-based stage play Night of 200 Billion Stars, but I wasn't. Rather I was delighted.

After a hundred years of Einstein and several years of encouragement from Bertrand Russell about "The Scientific Outlook", followed by the computer and internet, mobile phones and all, it would rather be surprising if somebody did not write a science-based play of the sort and get acclaimed for it. For science is becoming popular, particularly through space travel, as many mysteries of the universe (or a few of them!), and also wonders of the sub-atomic level are revealed, especially with the magical qualities of the quantum (or a few of them) thrown in. Many books have even been trying to claim that, because of the Minkowski equation of space with time from which they assume the existence of something called "curved space time", time travel is a "scientific possibility"; and although they are wrong because the Minkowski theory is false, the layman is bound to be intrigued.

It's not very clever describing science as just one of the many equally valid ways of looking at the world without scrutinising why, how and what the scientist believes or examines the world for. His job is the logical and systematic search of what is really there in nature and which can be employed to serve the life of man to make it comfortable, longer and happy; for although death awaits us all, scientists want to make human life longer, safer and more comfortable and happy--- by and large, the evils of science come from politicians (some of whom are the most devious of human beings), not the scientific researchers.[85]

---

[84] *History of Western Philosophy*, Book Three, Ch. 6.
[85] Of course, a few individual scientists may be devilish, but, on the whole, science is for the benefit of mankind not its destruction and even individual treachery is often perpetrated in extreme secrecy simply because the scientific community could destroy the traitor without mercy on the general

And one thing that the religions and the anti-science brigades forget is that science is progressive; we wrestle scientific or dependable knowledge form barren, even hostile, nature. Take scientific medicine as a prime example. Obviously it evolved. Man did not know a thing about scientific medicine when life began. It only gradually and even painfully evolved from the researches (and the research methods had to be learnt) of countless individuals, many of whom never lived to enjoy the fruits of their labour. So now that we have grown to know scientific medicine, go to the moon and beyond in search of more scientific knowledge to use for the improvement of life on earth (for instance, astronomy may seem to be a mere academic exercise, but it is from astronomy we are learning how to deflect or explode asteroids likely to end life on earth), we can now eradicate many diseases including polio and small pox;[86] we have also invented the computer, the internet and wireless communication. So, I repeat, it is not very clever to say science is but only one of many equally valid ways of looking at the world---it may be so, but only systematic, logical, scientific knowledge brings lasting benefits. In that sphere there is no rival to science. Besides, nobody knows what is 'the valid' way of looking at the world. Even scientists make no such claim. Their method is to use our most incisive organ (the eye) to observe and report what they see and use them for human salvation. They are not claiming to be capable of doing more than this---but it is enough. It gets us most of the things we need for normal life.

Next, I quote from the popular book, Human Situation, by Professor Macneile Dixon[87]: "In the great arch of night above our heads about five thousand stars may be seen by the naked eye. In their marchings and counter-marchings they make a brave show, yet are in fact scarcely so much as a swarm of bees in all Asia, a spray of blossom in the limitless abyss, where 'a hundred thousand million stars make one galaxy, and a hundred thousand million galaxies the universe'. The stars

---

understanding that science is for salvation.

[86] I say 'we can' do so; more optimistic experts claim that we have already achieved the total eradication of these terrible diseases---but I am still keeping my fingers cross that some terrorists will not be able to resurrect them.

[87] In its day this book was almost as popular as the Bible.

we see are but a handful, and their removal would not disturb by as much as a decimal the calculations of the angle of their courses. We may be sure that for every human being in the world there is not one star apiece---there are ten thousand. Viewed from the bodily angle, no comparisons can express the insignificance of man among the cosmic magnitudes upon which our astronomers exhaust their eloquence...The earth is a mote of dust, and the sun itself a diminutive firefly. We inhabit the puny satellite of an inferior orb. There are millions of stars so immense that room could be found for millions of our petty sun in one of them."[88] This was written nearly a hundred years ago. We have to update the figures. According to the Philip's Concise World Atlas, p.3, "At least a billion galaxies are scattered through the Universe, though the discoveries made by the Hubble Space Telescope suggest that there may be far more than once thought, and some estimates are as high as 100 billion. The largest galaxies contain trillions of stars, while small ones contain less than a billion." If we can train researchers to establish such complex and yet accurate details of the universe, then we can also train them to do the other parts of philosophers' work, for this is also philosophy.

Scientists, or astronomers specifically, can now speak with greater authority about the universe than philosophers. To put it another way, what used to be the exclusive preserve of technical philosophy, that of telling us the nature and composition of the universe, as a matter of speculation or inference, are now displayed openly in elementary books by astronomers with proofs or all the necessary physical evidence required.

For our third example of scientific truths, as opposed to philosophical speculations, on the understanding that man requires both for his material and intellectual advancement, I quote from the New Scientist: "QUANTUM electrodynamics is arguably the most successful scientific theory there has ever been. With stunning precision, it explains the interaction of electromagnetic radiation (including light) with electrons and other charged particles. It is on QED that quantum chromodynamics, the theory of the strong interaction, is modelled."[89] This is the solid ground upon which Quantum

---

[88] *The Human Situation*, Edward Arnold, London, 1937, p. 155.

mechanics is built. Two Nobel Prize winners put the same idea in different words. In case the Magazine presentation strikes any readers as down-market, I will presently quote them for their satisfaction. First, Professor Richard Feynman began chapter 4 of his book, QED, saying: "...I am going to talk about problems associated with the theory of quantum electrodynamics itself, supposing that all there is in the world is electrons and photons..." and the wonderful Louis de Brogle also said: "Without the Quanta was not anything made that was made."[90] Ordinarily we know light as immaterial. These statements are not only shocking but bother on the ridiculous. Yet they are true, and philosophers who argue against them rather make themselves ridiculous. The religions may object because they deliberately like to object to scientific progress as it undermines their beliefs and make them the laughing stock of modern man---and they will tell you that the more they are mocked the more they like it because it tells them that they are having some effects, whereas they know that religious talk should not be heard at all by people dealing with true knowledge. But a philosopher, as a man of learning, the lover of wisdom, cannot contradict science with mere assertions derived from his arm-chair conclusions. He must incorporate the discoveries of science in his thoughts. There is no way he could do that unless he learns to become a Philosopher/Scientist, which refers to somebody who reasons by means of scientific facts only. So let's go on to show examples of philosophical sayings that fail this test.

First, I quote from the Oxford philosopher Professor William Kneal's book On Having Mind (Cambridge, 1962.) In concluding a small book about how and why we have minds, he wrote: "We must retain the Platonic notion of mental events which are distinct from anything in the physical world and manifest a special kind of connectedness." So, according to this 'learned' professor, there is a non-physical world in addition to the physical one we live in, and because of that the Platonic notion of mental events (which deals with that mythical world), is the true theory of how and why we have minds.[91]

---

[89] From The *New Scientist*, 8th Jan. 1994.
[90] Louis de Broglie in Matter and Light, Allen & Unwin, London, 1939.
[91] By mental events he is not telling us about the minute, hidden and invisible causes which are also physical, that prop-up bulky matter. He is

In fact, the quantum theory undermines the Platonic idea. The invisible world is that of the quantum, namely images of things can be cut off if the lights from any object do not reach the eye. Bishop Berkeley has already proved this without even knowing it, and I will come to that in a moment. But his 'proof' allows us to infer that light radiation from objects conveys the images as the surface silhouettes of objects for us to see them, for we know that the particles of light, the photons, are naturally coloured. In an essay on Bishop Berkeley, in his monumental book History of western Philosophy (which I urge the reader to consult about this matter), Bertrand Russell wrote: "Berkeley advances valid arguments in favour of a certain important conclusion, though not quite in favour of the conclusion that he thinks he is proving. He thinks he is proving that all reality is mental; what he is proving is that we perceive qualities, not things, and that qualities are relative to the percipient."

My next example of how philosophers say things about reality to contradict (and therefore reject) what scientists actually find 'out there' is taken from a review of the book, "Science, Perception and Reality", by Professor Wilfred Sellars, an august Harvard professor of philosophy, a man of learning or of wisdom. Only it turns out that what he knows is scientifically rubbish:"Professor Sellars nowhere states the purpose of his book, and, since it is a collection of independent essays on a variety of topics, its intention can be judged only from its title. It is therefore not unfair to take it as an attempt at a philosophy of science…From this point of view, its value is slight. It reverts wholeheartedly to the Mill type of bypassing scientific activity, and analyses questions which are quite independent of anything scientists do. To take but one example: 'Philosophers have been fascinated by the fact that one can't have the concept of white without being able to see things as white, indeed, until one has actually seen something white. But this

---

rather bizarrely referring to something akin to magic in homage to Plato instead of observation. He is dreaming rather than looking at the world to see what is there. And he's so sure of himself that he insists that 'we must' do what he prescribes! With professors like this in our great institutions writing this kind of nonsense about philosophy, who can blame scientists for laughing at philosophers? Either we call what is done in philosophy departments 'logic', and restricted to logical studies or philosophers should pay attention to science rather than Plato's useless mysticism.

can be explained without assuming that sensation is a consciousness, for example, of white things as white". The reviewer, Professor Herbert Dingle comments, "A Scientist could scarcely care less for what has fascinated philosophers. He does not regard this as something to be explained. He starts with observation and forms concepts as required to express the relations he finds between them..."[92]

My third quotation is the withering criticism of philosophers by another philosopher (mentioned before), only this one is a follower of Bertrand Russell, Sir Karl Popper: "You see, the history of man is a queer thing. It's a history of a succession of attacks of intellectual madness, of all sorts of strange intellectual fashions. I don't need to give many examples of revolts against reason (such as Existentialism), for we know how strongly certain fashions have taken hold, not only just in a comparatively small insular group, but, in large parts of mankind. Russell saw these things in that light, and so do I...In the long history of philosophy there are many more philosophical arguments of which I feel ashamed than philosophical arguments of which I am proud...Yes, I cannot say I am proud of being called a philosopher..."[93]

I have quoted extracts from the works of scientists and philosophers. I believe it will not be difficult to decide which of these thinkers are speaking the truth, especially given the harsh criticism of philosophers by Sir Karl Popper, one of the great thinkers of the 20th century. Traditionally we are told that the purpose of philosophy is to think about the universe and the world we live in. If that thinking exercise is so bad that one of the leaders in the field is ashamed to be called a philosopher, then we have to agree that something is wrong with either philosophy or how we train our philosophers. I think both are misguided: The traditional topics discussed by philosophers are no longer relevant to the world we live in, and how we train them is also archaic. Of course, it is common knowledge that since Einstein and Bertrand Russell some institutions have

---

[92] Professor Herbert Dingle, reviewing *Science, Perception and Reality*, TLS 25th October, 1963.

[93] Sir Karl Popper, in conversation with Strawson, Warnock and Bryan Magee, published in *Modern British Philosophy*, London, Secker & Warburg, 1971. Sir Karl Popper was sending out a message because he knew that what he was saying would be published in a book.

started to call what they do as "The Philosophy of Science." But I am not convinced because when I browse through some philosophical journals I see that they are publishing articles on the same old traditional topics from Plato to Kant, with particular emphasis on the history of the subject without stressing what is right and what is wrong in the ideas of previous philosophers.

Yet that is what made Bertrand Russell's History of Western Philosophy particularly valuable. The old philosophies remain footnotes to Plato, by whom there are two worlds: the visible world of physics and the invisible realm science cannot reach, as Professor William Kneal put it, it consists of the world of "mental events which are distinct from anything in the physical world..." So mental events are to be preferred to actual physical events occurring out there and influencing our lives physically.[94] This is anathema to scientists who are already baffled by the quantum theory, which amounts to the analysis of matter down and down to the invisible sub-atomic particles. In fact, it does appear that material physics is finished; the physical analysis of matter is at an end.[95] There is no longer any solid matter to analyse as it has been realised that solid matter is composed of invisible matter down to nothingness, and all of them are known, except the Higgs boson which is still under active investigation.

Of course there is a problem with induction, as there are numerous problems with all aspects of life, especially in medicine. What scientists are saying is that they don't know everything; so many of the quandaries in life defy scientific explanation. But they don't allow them to frustrate scientific activity. They rather get on with it, and they are right because they get results. As Bertrand Russell has warned, in physics for instance, we have to obey scientists on pain of death, because whether traditional philosophy approves or not the scientific method can destroy life.

---

[94] What scientists are asking for is for the resources to go out there, see what is there and study it for solving human problems, as against philosophers telling them to honour the Platonic mysticism.

[95] There is no doubt, of course, that technological innovations will never end. The things that can be invented out of the particles known to exist are literally infinite. At present we need inventors more than mathematical theories in all branches of science.

On the contrary, instead of doing everything in accordance with what the traditional philosophers have said or are still telling us to do[96], Russell was bold and commendably judgemental; he condemned some thinkers; he praised others, almost precisely as Sir Karl Popper has also done, because Popper was one of the 'great' followers of Russell. It must be stressed that the notion that philosophy has to move closer to science has been known for years. Let me illustrate what I mean with a few quotations from Russell's great book. I remind the reader that we are discussing how Russell, guided by logic and history, science and common sense, praised or condemned some philosophers, and I regard that as one of the best things he did. It is extremely important to judge philosophical ideas by the requirements of human welfare and survival. For it is not disputed that philosophy is necessary; the contention is that, as the quotation from Professor Kneal has shown, not many of them respect scientific ideas or the scientific way of looking at the world and ordinary human welfare and what is necessary for human survival.

About John Mill Russell wrote: "John Stuart Mill, in his Utilitarianism, offers an argument which is so fallacious that it is hard to understand how he can have thought it valid."[97] On Aristotle he wrote: "In reading any important philosopher, but most of all in reading Aristotle, it is necessary to study him in two ways: with reference to his predecessors, and with reference to his successors. In the former aspect, Aristotle's merits are enormous. In the latter, his demerits are equally enormous. For his demerits, however, his successors are more responsible than he is." Thus, in spite of his many faults, especially in his Metaphysics, Aristotle got off lightly. But on Rousseau Russell was clear that he invented evil: "He is the

---

[96] This is the trouble with some religions. They do not want to move ahead but rather repeat what their ancient books are telling them to do, however ridiculous in our present modern world.

[97] Poor old Mill. One of the lessons I have learnt is that the modern view that philosophers merely debate topics as intellectual exercise using logic is not correct. I seem to sense that all philosophers want to be remembered for discovering or solving something, saying something intelligent. Thus why some of them ignore science is difficult to understand. For scientists also set out to solve problems, using observation so that what they find can be repeated by others. If we train scholars to combine the two disciplines collectively man will come to command a powerful mind.

father of the romantic movement. Hitler is an outcome of Rousseau, Roosevelt and Churchill, of Locke." To Russell, Nietzsche was also a merchant of evil, and who can blame him? He said: "I dislike Nietzsche because he likes the contemplation of pain, because he erects conceit into duty, because the men whom he most admires are conquerors, whose glory is cleverness in causing men to die." But Hobbes and Bacon came in for some praise. He wrote, "Hobbes (1588-1679) is a philosopher whom it is difficult to classify. He was an empiricist, like Locke, Berkeley, and Hume, but unlike them he was an admirer of mathematical method, not only pure mathematics, but in its applications." "Francis Bacon (1561-1626), although his philosophy is in many ways unsatisfactory, has permanent importance as the founder of modern inductive method and the pioneer in the attempt at logical systematization of scientific procedure." Also he said: "Spinoza (1632-77) is the noblest and most lovable of the great philosophers." On Schopenhauer he wrote, "Historically, two things are important about Schopenhauer: his pessimism and his doctrine that will is superior to knowledge. His pessimism made it possible for men to take to philosophy without having to persuade themselves that all evil can be explained away, and in this way, as an antidote, it was useful. From a scientific point of view, optimism and pessimism are alike objectionable: optimism assumes, or attempts to prove, that the universe exists to please us, and pessimism, that it exists to displease us. Scientifically, there is no evidence that it is concerned with us one way or the other." Lastly, let me conclude this section with his views about Rene Descartes, who else?[98] The man who made French intellectually almost as great as those of ancient Greece! Of course he is praised by Russell, and rightly so: "Rene Descartes (1596-1650) is usually considered the founder of modern philosophy, and, I think, rightly. He is the first man of high philosophic capacity whose outlook is

---

[98] Einstein was not the only philosopher/scientist. Aristotle, Descartes, Locke, Russell, Sir James Jeans, Professor Eddington, Professor A.N. Whitehead among others, were all philosopher/scientists. In my judgement Einstein was the most famous, Aristotle perhaps the greatest and Russell was the most popular. I believe it all goes back to Aristotle. After he taught us about the "logic of things", the cleverest minds realised that those studying the logic of things and those theorising about things in metaphysics ought to have ways of complementing each other's work.

profoundly affected by the new physics and astronomy. While it is true that he retains much of scholasticism, he does not accept foundations laid by predecessors, but endeavours to construct a complete philosophic edifice de novo. This had not happened since Aristotle, and is a sign of the new self-confidence that resulted from the progress of science." This is a description of a Philosopher/Scientist, not simply a thinker in the mould of Professor William Kneal and his 'mental event' or Professor Wilfred Sellars, who, according to Herbert Dingle, "...reverts wholeheartedly to the Mill type of bypassing of scientific activity."

Science is always good unless it is deliberately misused by some evil men, mostly for political purposes. Otherwise the basic aim of science is human salvation. However, as we have seen, philosophy is not always good, but very, very important because scientists cannot adequately think about the value of what they do; yet there are thousands of them; theories abound; some contradict others. It is necessary that a class of clever persons, or thinkers (suitably trained), makes it as part of its business to look at what scientists do overall and advise them in the human interest. I believe only scholars who philosophise can do that. They do not necessarily have to have doctorates from august universities, but they must think as philosophers not so much through speculations as through what scientists are finding out about people, the world and the universe at large.

Training, or education, makes a man. Lawyers coming out of law schools, as a prime example, always seem to have been brainwashed to be instinctively against crime, democratic, fair in their judgements about human frailty and dead against torture, unless they are basically evil in nature. Otherwise even when a person is found guilty of a heinous crime, his lawyer would plead mitigation and mercy for him or her. They even campaign for improvements in prison conditions. Why can't we train philosopher/Scientists as well---that is, to be equally fanatical about science in its noble pursuit of reliable truths in the physical world, medicine and society itself? To put it another way, why can't we devise a system of training to make competent people, men and women, capable of understanding science and instinctively think about it philosophically. Let me suggest how we could go about this training, on the understanding that I am a fallible human being and that my suggestions may not be the best imaginable. However, somebody has to start the ball rolling.

First, the philosophers: actually nobody can prescribe what must be taught to scholars to make them Philosopher/Scientists forever. The subjects will keep changing. But, on the whole, as we have seen, if philosophers concentrate on the Platonic mental events so that they can only write footnotes to Plato; or, conversely, if they chose the Mill type of bypassing scientific activity, they would have nothing of interest to say to scientists. Yet science dominates every aspect of human life and must be given competent intellectual, or philosophical, guidance. With this proviso, and roughly speaking, I think that for future philosophers to be able to understand science properly, they will have to be taught, among other things, subjects like logic and mathematics, metaphysics and astronomy, ethics and psychology, and also the general principles of scientific medicine. They cannot ignore history and literature, meaning the works of great writers and literary criticism.[99] All the sciences, particularly quantum physics, will have to be taught in philosophical classes. Bertrand Russell's books must be read, and read very well. Methods must be found for teaching Relativity, special and general. It must be taught without the Minkowski contribution; then, conversely, taught with it, so that scholarly will come to understand the 3+1 formula as opposed to the Minkowski purported equation of space with time to abolish the 3+1 system and contrast the differences. By showing students what is meant by merging space with time and why the Minkowski technique for doing so is untenable because of his use of imaginary time coordinates, they will come to understand that reality is based on observation, not on somebody's mathematics alone.[100] For if physical reality is 3+1,

---

[99] C. P. Snow's 'Two Cultures' should become a textbook. Writers like Noam Chomsky and linguistic philosophy should also be taught. Science is not a hobby. Science is not for scientists alone. It is for everybody's benefit. It is necessary to communicate it properly to all and sundry. Professor Sir Arthur Eddington, Sir James Jeans, Karl Popper and others including A J Ayer, A. J. P. Taylor and Professor Macneile Dixon and his *Human Situation* must all be taught for their intellectual achievements, their humanity and their literary merits.

mathematics cannot change it to one of four-dimensional continuum.[101] The above tentative suggestions may be found useful in planning the philosophical education of future generations of philosopher/scientists. Now let us look at the education of scientists to make them appreciate philosophy and philosophers.

The majority of scientists are woefully ignorant of the other branches of science. Also, it is they who are moaning that they cannot understand relativity without the Minkowski formula, which means that, since the Minkowski formula is logically flawed, they do not properly understand relativity. So the above training programme plus their own specialist fields will be required---not much else needs to be said on this subject. The only problem concerns the complex mathematical interpretations of general relativity based on the concept of "curved-space-time". By this monster time travel is said to be possible. Of course, if space is the same thing as time then it could be. But the fact is, if the Minkowski equation of space to time is false, then when space curves it cannot take time with it. The whole concept will have to be re-examined on the basis of the 3+1 formula because it means four-dimensional space or 4D geometry does not exist, yet scientists are propounding all their theories on the basis of four-dimensional space or space-time. It is a serious problem and I don't know how they are going to solve it, for so many theories since Einstein will have to be reviewed. In plain language, many scientists are going to have to try to understand relativity without the concept of 4-D geometry, or the idea that space and time constitute one entity---many scientific theories are going have to be discarded as happened after Einstein when the eather hypothesis was abandoned.

Unashamedly, of course, this chapter is entirely about the status of Einstein as a thinker---and God knows he was some thinker---and the changes in our intellectual traditions that have sprang up after him. But was he also a philosopher, perhaps even one of the greatest of all time?[102] We first must show what

---

[100] Thus they will also come to understand what is the coordinate system in modern physics.

[101] It creates discrepancy between the mathematical symbols and the essence of physical reality---my main objection to the Minkowski formula for equating space to time.

makes a philosopher. A philosopher is a professional thinker who has provided an insight about one or many of the aspects of nature and man that could not have been discovered from any other source. If his discovery is great, we call him one of the great philosophers. We know Einstein was a scientists and one of the greatest; but did he qualify by the above definition to be labelled a great philosopher?

Well, every reader can see that I adored him enormously. His greatest scientific achievement is call general relativity; in fact, it is entirely about gravity; and we regard his discovery as one of the greatest in science because it was so strange and yet true. We are all astounded and cannot imagine how he got his knowledge that gravity is caused by the curvature of space---although Riemannian geometry did help him a lot. But, strictly speaking, gravity is part of science. It is physics. It is not philosophy.

The philosophical aspect of his theory of gravity is the insight that the cosmos consists of two metrics: the special relativity frame and the general relativity frame. This was recognised at once as an extremely important original philosophy: two worlds in one universe. One in which life could exist and a second one where there is nowhere anywhere for life to evolve and flourish. So there could be no time there as well. For, as Russell has observed, time is a construction—sentience is required; and you have to have somewhere to live before you could undertake any kind of construction, physical as well as mental!

Without Einstein we could never have known that there is another metric in the universe (called general relativity) where life is virtually impossible. And since this has been confirmed, he becomes a great philosopher by that apparently simple observation or discovery. Yet even that is dwarfed by another momentous insight: the notion that the Lorentz concept of 'local time' is time. Lorentz thought it was a mathematical curiosity,

---

[102] Without doubt, Isaac Newton was a remarkable man, a great, original and revolutionary scientist, but he was conventional. With the normal training in physics, probably to the level of a doctorate, it was easy to understand him. Not so with Einstein. He changed the subject, first in special relativity with his notions of time, and also in general relativity with the first and only explanation for the cause of gravity, and with so many other ideas about the quantum and so forth that we are still struggling to understand him properly.

and put it aside. Later he was to admit that he failed to discover special relativity simply because he did not attach due importance to the fact that local time means somebody creating an alternative time as against absolute time, which was the insight Einstein needed to arrive at the idea that time can begin by anybody from anywhere and therefore it cannot be either absolute, fixed, or generally permeating the cosmos as a rigid entity; fixed, as many people believed, by God---all because it is so mysterious.

Thus Einstein more than qualified for the grand title of a great philosopher/scientist for that one idea alone---yet we all know that there was more to come, much more. For the point is, it is not the size of a tome that determines its value; not the complexity of an idea that we worship; not the intricate mathematics of any proposal we want to see. What makes any intellectual contribution important or even momentous is its value in our scheme of knowledge. Some tomes are valued mostly by paper manufacturers, and so heavy that they could be lethal missiles in gang warfare. One example is a book published by Cambridge University Press in 1995. A major tome by an incredibly high-brow team of international professors and notable scholars. It was entitled "Time's Arrows Today: Recent physical and philosophical Work on the Direction of Time". Yet a few years later Professor Yourgrau wrote his book about the forgotten legacy of Godel and Einstein concerning the fact that time cannot exist under relativity. He said he was telling us about the "revolutionary notion of a world without time." So who can we believe? The mighty team of international scholars' thesis about the direction of time (something they know to be definitely in existence), or the legacy of Godel and Einstein to the effect that time could not exist under relativity? Only a brave philosopher/scientist would dare to hazard an opinion; no one else should dare to intervene. One side or the other is wrong; yet we know that one side is so powerful that if it got angry with any amateurish (non mathematical) intervention, some people could lose their jobs!

As a retired diplomat in my seventies, I have no job to lose.[103] So I dare to state that, technically there is no time if each 'body'

---

[103] We Africans have a proverb, which translates as 'dead sheep is not afraid of the knife!'

has to have its own time so that "there are as many times as there are inertial bodies". If any inertial body has not invented its time, it will of course have no time until it has created one. It means to have time sentience is required. Somebody must be there to invent the time---or place the points for the year and pare it down to the seconds. However, we know that, thanks to Einstein, this applies to quantified time only. Otherwise things grow in the absence of sentience; and it takes time for things to grow. Therefore there is natural time---natural periods during which things, events, action and growth occur. What we add, the human contribution, is the process of reducing it to culturally manageable units, using repetitive cycles. Given these facts, the mighty team of international thinkers failed because they knew nothing of 'quantified time'. They just talk about time; and Professor Yourgrau is right, for time in that ancient sense linked to motion does not seem to exist, since there is not only one form of motion but billions, and they do not all pass by in tandem---i.e. there can be no general passage of existence because existence is multitudinous, and we all, billions of us, see the world differently. Only deliberately 'constructed' or 'Quantified' time (human in origin) unites us all.[104]

On the other hand, there is time in society. Wherever there are 'Beings' there will be time, for man cannot live successively without time for long. What to do during the day, and what to do at night require time for regulation; similarly, what to do during the coming days, weeks, months and year, requires time for planning. Hence quantified time, being an insight borne of the combination of scientific facts and philosophic intuition. We need many more of such thinkers especially as society grows more and more complex; and while training cannot produce anybody half as good as an original one like Einstein at least we can try.

---

[104] It is true that time does not exist as a recognisable physical entity but chemical processes do take time or involve waiting---which is what we call time. The challenge is to define that process we call and use as time in both its psychological and physical aspects. The earth's journey round the sun is, of course, physical and we use it as time. The internal knowledge of the lengths of time units to be able to tell that a year is longer than a months, is the psychological aspects of time---both need interpretations, or a theory that can account for both---because they do occur and therefore do exist.

4

WHAT IS SPACE-TIME

The technical definition of space-time is stated as "A mathematical space specified by three space coordinates and one of time." With this definition it is claimed that space and time become one entity. It is argued here that the notion is wrong and that, therefore, there is no such thing as 4-D geometry or four-dimensional space continuum since the Minkowski formula for it relies on imaginary time coordinates. My earnest plea is that if physicists are to become capable of solving their problems without waiting (for ever) for another Einstein, then they should take a good look at the concept of four-dimensional continuum. Even Eddington said it may be used but it should not be forgotten that it is nevertheless arbitrary and fictitious. The Minkowski formula is not wrong because of the mathematical deductions or the incorporation of c, as in the equation $\sqrt{-1}.ct$... It is wrong because of the imaginary time coordinates upon which the subsequent mathematical deductions are based, and represented by i.

Let me begin with a controversial assertion, controversial but true, namely everybody writing or thinking about time today is making five mistakes: (1) the first is that they all forget that there is no longer a universal time so that the time they are discussing appears to be plainly universal; it gives the impression that it exists somewhere.[105] Let me stress that there is no longer a time system that permeates the whole universe. Under relativity, every world has to invent its own time; so you must begin by showing how earth time was invented, created or came about, as it was not 'given' by God.(2) They also assume that there is such a thing as four-dimensional continuum, or 4-D geometry as Minkowski supposed with his mathematics alone without physical proof. Nobody in recent times has been brave enough to suggest even tentatively that 4-G geometry may not exist. Yet Bertrand Russell and Professor Sir Arthur Eddington described the Minkowski theory as arbitrary, although Eddington said it may be used, but people should not forget that it is nevertheless arbitrary and fictitious. Why, therefore, is the term 'space-time' still in use in science without qualification, or without mentioning the caveat of Professor Eddington?[106] (3) Nobody ever tries to define time; people just use the clock.[107] (4) In all cases time is assumed to be naturally existing; or they speak about it as if it is the same thing as 'Being', just being there or motion, the general passage of existence. But there is no such general motion. All things are either static at the visual level or they are in metamorphosis; they are not journeying to anywhere. (5) Everybody is interested in time travel, and most of the writers on time believe that it is feasible. As a result, most of their theories are wrong. By rejecting 1-5 above, I was able to

---

[105] But time 'constructed' on this planet does not exist anywhere; we can only notice how it is passing through nature with regular or cyclical motions. We cannot know time in any other way after relativity. That is why I chose Russell's statement that 'There is no longer a universal time' as one of the postulates upon which to build my theory.

[106] It was the merging of space and time (already evident in special relativity as Bertrand Russell has pointed out), that Minkowski said he was rendering in geometry; but Einstein himself never called it 'space-time'. Now there is a problem with the term even if we reject it. Scientists are so fond of it that we may have to retain it but with a new meaning---that it does not mean space and time constitute one entity, only that time cannot be created without space. "Relation between points" implies space.

[107] Even philosophers have found it daunting. Very few of them discuss it in depth. Scientists are happy to hide behind the Minkowski complex mathematics, knowing it is not true!

conceive my explanation for the nature and passage of time (outlined in this book) without much difficult.

The reader may wish to know why 'Being', or just being there, is often taken to be time or can be regarded as time going.[108] I will do my best to explain what it means. The reason is that you age; you are just being there but you are ageing. When you age it is either time moving you on, or your time is moving on in the religious sense. Either case, it is vital for human existence or the definition of life. It shows that time is closely associated with life; and since it cannot be stopped or avoided, it seems to be life itself and needs to be clearly defined, that is why philosophy and religion are involved. For when you define time you are indirectly defining what makes life what it is. To assume that it is imaginary and decide the nature of physical reality by that assumption is intellectually an abnegation of duty. It just is not very clever.

It is in this sense that the concept of space-time becomes vital for human existence. It is not only physics or an insignificant section of mathematics. It is rather part of the metaphysical criteria for existence, namely is the space in which we live the same thing as the time by which we live or not? Space-time is supposed to be 4-D geometry or four dimensional space incorporating time in space so that the mathematics for representing physical reality can be written simply as 'S=ct...', meaning that space is the same thing as time observed through the medium of light---very technical, but all to the liking of mathematicians who like to be regarded as angels of the human intellect set apart from the common run of mankind.[109]

---

[108] *See the chapter on "Time and Quantified Time" below. You cannot mechanise the sort of time implied in just being there for the clock. That is to say, you cannot put it in the clock to tick for your use in society or culture. So, in the end, to know how much you have aged while just being there, you have to use quantified time. And this is where things begin to get more complicated, because quantified time (whether by atomic pulses or the earth's orbits of the sun) can only show how time is passing ---therefore we can only know how time is passing and never what it is. By the concept of 'Quantified time', time is given a very credible interpretation that is compatible with relativity, logic, physics and ordinary commonsense. As such, it is superior to all concepts of time.*

[109] We all know the idea that time is 'naturally' incorporated in (or only mathematically incorporated 'into') space was used in the field equations of

However, where or how can one apprehend space-time? This is important because we do not live in the minds of mathematicians.[110] Furthermore, space-time is supposed to be part of science, or, in fact, the very foundation of physical reality in science overall, that is the reason the term pops up in every scientific statement about reality. On the other hand, all science is based on concrete physical evidence, so much so that, however abstract, every scientific idea must be traceable to something in physical existence, otherwise it is not better than the discredited, speculative philosophies of the metaphysicians and should be thrown out.

In answer to this simple, basic question, we are confidently and eruditely (even pompously) informed that, for those of us ignorant of counterintuitive mathematics, space-time is "a mathematical space [physically so as to qualify as a scientific proposal] specified by three space coordinates and one of time." With this definition any space is time, or space-time. The three space coordinates are obvious enough, but the time coordinate is taken as 'given' by scientists of all people. We are not allowed to ask how it is given. If you press, they tell you that it is an imaginary time---which displeased Bertrand Russell most profoundly.[111]

Einstein's general relativity. Because of that, people who are not very conversant with the process of philosophical thinking assume that it must be true or logically valid. On the other hand, we now learn from David Hilbert that Einstein never even understood it. But of course he understood it, otherwise how could he have known where to place it? I can only speculate that Einstein used it to placate his mathematical critics, knowing it was logically flawed, but also knew that it could in no way affect the truth of his general relativity proposals, whether or not time is independent of space. As Russell has pointed out, the concept is chiefly for the convenience of mathematicians---it helps them to define space, time and physical reality in one equation as 's=ct...' But this is almost fraudulent, because, logically, whether space is the same as time or time is the same as space, the second is always the same---and that is what matters in the study of time. So you could say Einstein played a more sophisticated game with his critics. A man like Einstein, as we are now told by David Hilbert, could never have missed the glaring logical anomaly in the Minkowski formula for equating space to time.

[111] See Ch. XXXVIII of Russell's Analysis of Matter.

Yet space-time is supposed to be time (or the time) under relativity but, because of the abstract nature of relativity, it is fused with space---hence the term. If this contraption (or space-time) is or can only be defined---and brought into physical reality--- by three coordinates of space and one of time, then where is that time coming from since time that is not absolute or generally permeating the cosmos as 'given' requires space to create?

To overcome this handicap the time is assumed to be imaginary. In his book "Relativity" Einstein put it beautifully: "...we must replace the usual time co-ordinate t by an imaginary magnitude $\sqrt{-1}.ct$ proportional to it..." Brilliant, Albert. He even went on to praise Minkowski. Like all converts, Einstein was showing more enthusiasm than his critics. Or probably a pretence to fool his critics. Somebody must dig deeper about his motives for this sham, because it was enough to send the ablest philosophers into intellectual convulsions. At once Bertrand Russell declared space-time to be arbitrary and logically untenable because it is based on imaginary time coordinates, and anything imaginary, either of time or whatever, has no place in science---or a discipline strictly based on concrete evidence as physics must be otherwise it loses the right to be feared as something that must be obeyed on pain of death. In other words, that is not what has made physics so fearsomely accurate and effective. Philosophy is individualistic, even sometimes idiosyncratic and silly; but physics is universally based on physical reality and must be followed strictly so that, once correctly invoked anywhere, it will work as well as everywhere else. If this rule is breached, physics becomes powerless; yet it is true that knowledge is power that is the reason physics is so fearsome, for it does, indeed, when correctly followed, invoke phenomenal powers over mankind and his environment. It cannot change nature, but it sure can show how to make it work effectively for good or ill. In the past, it has been used mostly for wars, as stupid and ignorant mankind sought to eliminate one another; but since the electronic revolution, we have become aware of its bountiful resources.

Therefore, Professor Sir Arthur Eddington, the founder of astrophysics, in a noble attempt to rescue the theory, issued a defiant caveat: "...we shall continue to employ it; but we must endeavour not to lose sight of its arbitrary and fictitious nature."[112] ---which is nonsense. An arbitrary theory cannot be useful in science. It is true that mathematicians tend to be mystical in their attitudes, but what is useful in mathematics (as a subject) is always related to observed fact. I therefore declare that the space-time concept is a grotesque distortion of relativity and physics in general and must now be abandoned. Courage is required, because, as the Americans said in the Deep South/Wild West, when the myth becomes reality, stick to the myth in the name of familiarity; it might have been harmless in their peculiar circumstances then, but space-time seems (specifically) to have been invented (as a religious or 'Communist' plot!) to counter the secular nature of Einstein's theory of time.

The pain in the hearts of all logicians and philosophers of rational bent is that the momentous implications of secular time are being neglected; no one can ever know what discoveries might have come to light if these implications were noted earlier; but what is clear is that they are so serious that even scientists want to run back to Newtonian absolute time for safety. Let ne repeat that, definitely Einstein divided the universe into two. One is where the laws of physics apply. It includes astronomical objects similar to our own in content. The other is where the elements we need to live and flourish are not present and is called the metric of general relativity. It is assumed that this second uninhabitable world is in the process of conversion, giving many university dons a field for innumerable assumptions, opinions and theories. 'Black hole' is the term used to describe this strange world where there is continual acceleration of matter towards infinite condensation, and so forth, but most of their speculations seem to be puerile.

It is not suggested that the universities and learned institutions have failed to attract brilliant minds in their service. They have. But they have wasted their talents working on the assumption that the world and universe is one of 4-D geometry and therefore, as Minkowski suggested, space and time constitute one entity. It should be clearly understood that there is a world of difference between working on the assumption of 4-D geometry or four-dimensional space, and on the contrary,

---

[112] See his *Mathematical Theory of Relativity*, Ch.1.1.

working on the 3+1 system of continuum. So, as CERN has now admitted, recent events in physics would seem to suggest that the notion of four-dimensional space will have to be revisited. (And let me repeat that not for the breach of c whether that is true or not, but for the i in Minkowski's ict equation.) This is a clever way of confessing or admitting guilt (by the CERN), of saying it might not be true of the world. But it will do. The problem is formulaic thinking. That means researching on the basis of established formulas. This was clearly illustrated in the eather debacle, but the lessons were not learnt.[113] Theories are always piling on formulas that are flawed, leading eventually to a revolution in the subject concerned.

When the Minkowski formula for equating space to time is rejected, all reality is changed---to what it might be no one knows because it has not yet been thoroughly analysed. However, it is enough to earn Einstein the rare title of 'Philosopher/Scientist' on a par with Pythagoras but much more on the secular side of mankind, which is another unique credit. Mathematicians may not like it; like Pythagoras, they tend to lean to the side of mysticism. Yet it was not the fault of Einstein that time is different in different places. As even Minkowski hinted in his original Cologne lecture about space-time, it came 'from the soils of experimental physics'. H. A. Lorentz found, accidentally, that time is different under different conditions. This implies that one hour here is not the same everywhere else. Thus, mankind had to try and discover how his 'seconds' are created; and with time being so fundamental to human nature, all reality was changed. It turned out that, logically, we can only get our seconds with the application of points to space, but sentience is required---somebody must be there to set or identify the points for the repetitive motions, or there will be no years and no seconds derived as fractions of the year,

---

[113] If a few of all those claiming to be as brilliant as Einstein paid a little critical attention to the formulas they build on, physics would be in a better shape today. As it is, many of their suppositions are flawed and therefore physics is shaky, although they pretend otherwise. Actually I think physics would have collapsed with ignominy years ago but for the fact that time is always there, so that however they write it out, it is there to respond---the main reason Einstein got away with using the Minkowski formula in the field equations of general relativity.

etc. Hence, Russell defined time as "Relation between points". Since these points can only be applied to space, it made the Minkowski formula for equating space to time completely illogical. To continue to theorise that 's=ct...' is therefore a distortion of physics.

THE ROLE OF TIME IN HISTORY

Normally, history is recorded from the point of view of time, showing how time is marching on, say, from the 1st century to the present and marching on to where exactly? Heaven or the end of the world? In reality, history is not the march of time, year after year, but the march of events. The dates are merely associated to events as the dates of occurrence. Our time is necessarily discrete; it cannot march. Only events have antecedents and consequences that go on and on forever as the story of human existence. From the dawn of existence these events can be traced continuously to the present and still going on.

History, as the march of time (including past, present and future), is the source of most myths and legends about time. The writers of such theories get away with it by carefully avoiding any logical definition of time. Before Einstein showed that cosmic or absolute time does not exist, and the great Bertrand Russell asked what then is measured by the clock, nobody could offer any logical or rational explanation of time except that it permeates all through nature and is the same everywhere, thanking Newton for his views about absolute time and absolute space; and it is true that Before Einstein Newton's ideas were unchallengeable. He was also loved because what he could not explain he attributed to God. But once absolute time and cosmic time were shown to be false and we were left with the task of finding the source of our time---and what it is--- the trail went cold. There is nothing to analyse. What we call time consists of repetitive cycles. We count the orbits of the sun as 'years', and pare the year down to the seconds or even the atomic units of time. What else is there to call 'time'? Yet that is passing time as explained above, and as I hope physicists who rely on the Minkowski four-dimensional space will soon find out.

All this adds up to make time necessarily discrete, since it consists of units of time beginning with the year, or a similar basic unit, pared down (for convenience) to the fractions as sub-units of the year---being the familiar seconds, hours and months, etc. Discrete time cannot march through nature.[114] In Amy view history is not the march of time but the march of events. Time has no history.[115] We remember yesterday through connected events not through connected time; the story of all history is similar. It is always the story of what happened to people, not what happened to the hours and minutes, or how the days came and went. The units of time are not different one day from another; they tell no story, only events do. For example, in geology time is secondary, as it should be. Instead, we trace the history of the earth since its

---

[114] Discrete time makes time travel a laughable proposition. The year 3000, for instance, is not existing anywhere ahead to be reached by time travel because if for any reason the earth fails to orbit the sun regularly till the year 3000, then that year will not come to pass, so reaching what does not exist can truly be described as 'a religious myth' that no rational thinker (scientist, philosopher or logician) should consider seriously. It is true that 'folding space' attributable to the curvature of space can curtail distances in space. However, if the Minkowski formula for equating space to time is fictitious and arbitrary then when space curves it cannot take time with it.

[115] The Einstein notion of time was an invitation to ponder seriously the nature and role of time in human affairs, since it affects the nature and concepts of physical reality. It is for this reason that the Minkowski distortion is seen as philosophically painful or hurtful. If true, reality is changed forever. However, since it is based on imaginary time coordinates, and therefore cannot be true in nature, we can all breath again. No mathematics can save it. It is beautiful but false and a grotesque distortion of physics, cosmology and part of astronomy.

formation from causative connections and consequences (what caused what?) That has nothing to do with time, but all to do with the deterministic metamorphosis of rocks and minerals. And that, properly, is the history of the earth---all history is the same. Historians and philosophers of history need to take note of this. To argue that at some time in the past there was such a thing as 'Time Zero', since when time has been marching on to provide the story of history from that 'date' to the present day, is nonsense (and probably religious nonsense.) Only events create history, for history is a story and only event can paint a story.

Of course, time is important but it is passing by because the units of time multiply to make time move on; this is deceptive; it makes history seem to be the story of time marching on. But I have already explained how we get the centuries: two years means two previous orbits of the sun, and so forth. At present historians record history from the point of view of time---what happened during days, weeks, months and years. That is wrong, and is the reason time seems to determine events. That historical tradition is unworthy of the modern (rational) man of science. It came from man's primitive ancestry when life was seen as 'Created' from a certain 'Day', in all religions, and has been moving on in time ever since 'that date or day'. In reality, scientifically and logically, events alone have antecedents and consequences that can go on and on and on to give a story; and history is always a story. The times and dates are merely associated with events to show when they occurred as a matter of cultural and legal necessity.

Suggestions from philosophy are serious because if true a lot of things in life are changed. So let me give instances of how history is not the march of time but of events. First of all, I must repeat for emphasis that history is a story; all history is a story; every history is a story. History is how anything happened or how some specific events occurred and what were the consequences; history is not the way time is supposed to be marching through nature, for after all, discrete time as we create with the earth-year cannot march; it merely replicates to continue. Let us illustrate this notion of history with the story of one country we are all familiar with---the United States. We all know how the United States came about; how America fought its way out of the British Empire to become an independent republic: one incident led to another until the war of independence broke out and the Americans won. Similarly, everybody will be able to trace the history (or the story) of how any country came about. For example, how India broke up to become three separate nations after independence: India, Pakistan and Bangladesh. The dates of these events are incidental to them because the events were not caused by the dates, not caused by time.[116] Time had nothing to do with the fact that Hindus and Muslims were unwilling to live as one nation in an independent India; neither did time cause the Pakistani corruption, neglect and incompetence that led East Bengal to break away and become Bangladesh.

---

[116] Time causes nothing on its own. Chemistry is always part of all causes; it may even be that chemistry causes the delay (waiting or period) we know as time. We are often told that time is the best healer; maybe that was the case before the rise of science. Today we think differently. The brain/body system causes healing---with the brain playing the bigger role—not 'over time', but rather it takes time to process the healing (the term 'mind' is avoided to outwit the religions). Even in personal, political and international relations, people mature to heal their own physical and emotional wounds through atomic processing. A. N. Whitehead is not given sufficient credit for his book about "Process and Reality". Yet, in my opinion, 'the truth' is known. All the answers to life's mysteries, in so far as we can discover them, are already buried in the works of Einstein, Bertrand Russell and Professor Whitehead. We are not aware of this because our present-day academics are not up to the job of interpreting their works for the general public. Instead we find them vying for media contracts to make money with old and cheap popular stories and myths of the religions and dead poets.

PAST, PRESENT AND FUTURE

These terms are part of the legends of time. From the sun's point of view there is no past, present and future. You would see vast space, some belonging to what we call 'past', others to 'the present', and where the earth is heading as 'the future'. So the terms must not be used to frustrate the rational definition of time. If time is not marching through nature then it is not moving from the past to the present and to the future; and we now know that, scientifically, time is no longer a universal entity marching through nature. So the past becomes aspects of our memory; the present is now; and the future is mere expectation, speculation or imagination.

It is well known that Einstein said past, present and future as aspects of time are illusions. He alone was entitled to make such comments without arguments; the rest of us are obliged to argue the case. Perhaps his comments would inspire us to search for the logical grounds to justify his views; all the same, we must advance arguments. He had earned the right to pontificate. In my opinion, they are incidental aspects of how the brain works, being memory (the capacity to recall as 'the past'), factual matters or the present, and the consequences of what has happened carried into the future. These concepts are based on the day and night system; that is the reason we have 'now' as the present, 'the past' (which is yesterday and everything that has happened in the past), and the sure knowledge that 'tomorrow' will come because we are alive and Day will follow Night as usual. Most quandaries of time began from our primitive past and can't stand up to logical scrutiny.

Often the International Date Lines are mentioned as showing the past and the present, with the future ahead of us. However, the truth is that we live on a planet whirling through interstellar space at the rate of about 20 miles a second, so fast that only the massive size of the earth in comparison with our tiny bodily structures, saves us from feeling uncomfortably dizzy due to the speed of the earth through space. In addition, things look different from different positions in space. From the sun's point of view, as I have said, what we regard as past, present and future will not be recognised. Rather the view will be a wide-open space; the fact that we divide it into past, present and future will not be known from the wide-open sprawl in the view from the sun as it is far above the earth. If you stand on mount Everest you will see wide open space; the fact that parts belong to different sovereign states is legally meaningless from that height---until you get down and try to squat somewhere in the full view of the legal sharks.

The planet itself, in comparison to the cosmos, is so minute that it produces no effects of any recognisable kind; so we seem to be on our own, tiny, ineffectual and lonely with only our religious myths for comfort. However, what happens on earth has antecedents and consequences that may go on and on and on indefinitely---at least some of them will. The fact that we divide these consequences of our actions and other events on earth (by means of International Date Lines) into past, present and future relates entirely to what happens on earth. The Date Lines may show 'space ahead', but that has nothing to do with us; there are no events there, and there is no guarantee we will even get there. On the other hand, we carry the consequences of past events (as unavoidable baggage) with us to the future.

We are told that the future is unknown, but that is not entirely correct. Most of the future is hidden in today's events that is why we can predict the future, sometimes with uncanny accuracy---which, of course, is not magic; it is clever logical deduction based on past and present events, simply because events do have consequences; that is the stuff of history. Thus I keep reminding the reader that history is not the march of time, but the march of successive events derived from past and present circumstances. By this interpretation of history (less a theory than matter of fact) religion is taken out of the stuff of history---what happened in history is what human beings experienced and its consequences that spill over to successive days; they are not caused by time but happen to occur at certain dates and times.

To my mind, these episodic experiences (caused by both the mystery of life and/or the effects of the features of the planet we live on, and against which we are utterly helpless),[117] are not time but conjunctional experiences of life, simply because they are not capable of being mechanised in a clock—yet, time, as defined here, is quantified time, such that it can be mechanised in a clock for general use, especially in science and rational thought. On the other hand, democratically speaking, what some poets and religious people call time in their heads and mythical suppositions is their own business.[118] The time we are discussing here is one that can be mechanised for general use on this planet. Otherwise if past, present and future are considered part of the physical reality of time what are they in dreams---what is their metaphysical status in dreams? The fact is that, like yesterday or tomorrow, the past, present and future divisions of nature by man arise from how we live, the revolutions of the earth and more importantly how the mind works, especially the memory mechanism, called "the capacity to recall or repeat."

---

[117] They appear to be part of physical reality, like dreams and hallucinations. However, in fact, they are some of the mysteries produced by the human mind in its reactions to natural causes and events—especially dreams and memory.

[118] The 'Two Cultures' syndrome is very much alive but is slightly different between philosophers and the television charlatans of the electronic age. Those who understand science are not good in philosophy; they regard it as useless due to endless arguments; and those who know some philosophy are shallow and useless in science. They say $E=MC^2$ is part of relativity and that it created 'The Bomb'---both are false. The all round scholars and lovers of wisdom have been starved and refereed out of existence by the money-grabbing media gurus and greedy devotees of television.

First, how we live. Man lives from day to day, or from one day to another. The last day is the past; the expected 'next day' is his future. However, it is wrong to consider these as part of the theory of time. They are not part of the furniture of the world; they occur to man and his way of living only---and that way of life may be right or wrong in nature. Time, on the other hand, is counting the orbits of the sun as years, and dividing the year down to seconds for cultural purposes. Secondly, how our mind works. Man is cursed and blessed with memory, so that we know of something called 'yesterday' or 'history'---but what has that to do with the nature or the interpretation of time? We think in that mould due entirely to how our mind works---which, again, may be right or wrong, but we cannot help it.

Hence the traditional divisions of time into past, present and future should never be considered part of the rational discussions of time in logic or rational thought. It is rather the product of ignorance in those days when science and rational thought were yet to be discovered. It causes unnecessary confusions but there is nothing we can do about it. However, it is not much of a problem, since we can choose to ignore all those theories about past, present and future as aspects of time.

The real problem for mankind is that the nature of time cannot be known and yet we have something we call 'time'. We know the year as the passage of time, for, indeed, it is the passage of time; so that we count the centuries from the number of passing years. Yet the years consist of repetitive cycles or motions---they are mere physical events. So let me stress that what we call time is the physical passage of human contraptions or mechanisms for recording time. Thus all we know is how time is passing, or how it passes through nature, because, lacking any idea as to what it really is, we use mere repetitive cycles and count them as 'years', and say the passing years constitute the passage of time.

Since this is all we know as time, questions about past, present and future belong to the study of process and reality, not time. Ultimately, they relate to the causative strings of events--- backwards and forwards, meaning what caused what and what the consequences are, or are going to be, using guesswork and previous experience with absolutely no guarantee of success. The mesh of theories and myths, contradictory notions and prognostications about past, present and future have been concocted by some poets and religious writers for their own doctrines and creeds most of which are used to 'prove' that time travel (and therefore religious beliefs about reincarnation) are necessary truths. Unscrupulous people are making money from them; others just desire the accolade of being God's spokespersons. Because of the use of language, human beings are cursed to endure such mental tortures from largely superstitious people---but no one should confuse them with the rational study of time for scientific and logical purposes.

Of course, it has been noted of late that some 'religious' scientists have been giving support to the poets and faith healers. For example, once again I remind the reader of what Professor Palle Yourgrau wrote to show what I mean: "Godel, the union of Einstein and Kafka, had for the first time in human history proved, from the equations of relativity, that time travel was not a philosopher's fantasy but a scientific possibility".(from "A World Without Time...", Ch. 1.) Sadly, there are no such equations in relativity. What there is comes from Herman Minkowski regarding 4-D geometry which we know as 'space-time' or four-dimensional space. It proved nothing about time travel because the formula is 'arbitrary and fictitious'. Mathematicians cling to it because they think that, from their point of view, it makes the understanding of relativity easy. As a result, I personally believe that the world has not yet completely understood relativity or what it tells us about the nature of time.

Several books about time travel have been published since Minkowski (all of them basing their theories on his arbitrary and fictitious 4-D geometry), and they have been very successful, which does not mean that they are logically defensible, but rather because man is basically gullible when it comes to mystical ideas about the universe and time, reincarnation and afterlife, time travel and God.

I think H.G. Wells bears much of the blame if not all of it. His 800,000 years time travel ahead could never have existed in any world, or on any planet, since the 800,000 years could only come from 800,000 orbits of the sun by the earth and therefore could not be in existence ahead to be reached through time travel. For the truth is that the years do not even exist. We call a number of orbits 'years', or one orbit 'one year', but they are actually the physical orbits of the sun.[119] How long it takes to orbit the sun is our basic unit of time and therefore cannot be logically defined in temporal terms. All other units of time are fractions of the year and can be defined in reference to the year. But we can never define the year as time, because it is merely a physical journey round the sun. However, we have nothing reliable to use for time, so we call one orbit 'a year' to represent it in our peculiar human consciousness of the rate of the passage of time.

Time, after all, is always in the past—i.e. after the orbit of the sun as a year--- because we can only know how time is passing. However, it is interesting that Wells was actually preceded by the Spanish Dramatist, Enrique Gaspar, who was cleverer because he concluded his play about time travel, El Anacronopete, with the confession that his story was only a dream. Quite right, too. Logically it could never happen in physical reality---the year is not ahead. It comes into existence only after the orbit of the sun.[120] So it is always in the past,

---

[119] It is a peculiar custom of mankind to count the orbits as 'years' or our basic units of time, because we have nothing else to use for the purpose. The year is then subdivided down to the familiar fractions with mathematics in conjunction with astronomy, or features of the planet. In the absence of that numerical time, which is human, what we get is just the passage of existence through consciousness in a bland and useless fashion---e.g. just sitting there and continually ageing and dying from old age is not an illustration of time to be used for the scientific study of time. It is mere chemistry. Thus it can be said that all human achievements are due to our having time to use, and it was 'constructed' by man.

[120] Like all units of time in discrete time, once the unit occurs, it is gone. The year is not reached till 31st December, but then it is gone; and another year begins. So we know the year in the past, or after it has passed. Therefore Well's 800,000 years have not yet been created, and how could anybody travel to or by what does not exist? Also, let me stress again that so long as we use the year as one unit of time, and more so, since we can only derive all other units as subdivisions of the year, our time is necessarily discrete. Atomic time is not different, only smaller in units. Some quite serious scientific thinkers believe that units of time (like the year or atomic units) exist independently and we travel to reach them. In fact, the journey round the sun is the time or what we call 'time'. One year is how long it takes to journey round the sun. Mathematical ingenuity (based on past experience), enables us to plan activities with time backwards and forwards, otherwise the time itself is created before, and known only in the past. If the journey round the sun is not completed, there will be no year and therefore no seconds

making time travel a clever and interesting literary game but scarcely the kind of rational notion for philosophers to take seriously---call it fantasy, or a dream.

The Spanish playwright was right. What has become increasingly evident since Darwin ended the role of God in existence is that people hope death is not the end, and time travel seems to them the surest evidence of that. Pythagoras has a lot to answer for. Furthermore, time travel to revisit the past is nonsense, for the time is known only in the past. Revisiting past episodes or events is another matter, or another way of dreaming that the people and materials involved are still there as they were experienced---for instance, that rivers, cloud formations, and weather conditions are all reversed into the state you saw them. Yet meanwhile some rivers will have dried up, especially in the deserts---the weather will have changed, clouds will have dispersed; also inevitably the events will have been carried forward and no longer existing in the past.

Whatever maybe the responsibility of H.G. Wells (alone or with his ilk),[121] it seems to me that even reputable publishers have lost their critical acumen and are blindly outbidding each other to grab manuscripts about time travel for commercial reasons as they can see that people are afraid of death and want to know about time travel. The fear of man about death seen as the ultimate end of life runs deep. In fact, if the Minkowski equations of space to time is arbitrary and fictitious, as Professor Sir Arthur Eddington tells us, then time travel can never be a scientific possibility and remains, not only a philosopher's fantasy, but also a religious superstition going all the way back to Pythagoras and the Transmigration of Souls.[122]

---

derived from the year as sub-units thereof.

[121] It may very well be that Wells (a great imaginative writer, scholar and thinker) also realised that time travel is something of a dream but was not honest enough to admit it.

[122] "J.W. Dunne's An Experiment with Time (1927) caused a sensation when first published. It proposed a concept of time in which time travel seemed possible. Max Planck could not fault Dunne's maths but said that his premises were incapable of proof and so were unscientific. Reverse Time Travel (1995), by the scientist B. Chapman, explores the subject in depth and, while not ruling out time travel, makes it sound pretty unattractive." (The Ultimate Book of Notes & Queries, The Guardian, Atlantic Books, London, 2002, Ch. 5, p 251.) I cannot go into this thorny debate. I merely

The real scientific idea about time travel is the folding of space due to the curvature of space and, therefore, gravity. That is no longer a theory. It came from Einstein's ideas and it happens to be the only theory in general relativity that has been confirmed. Sir Arthur Eddington was leader of the scientific team that confirmed it, so he knew what he was talking about when he described the Minkowski formula based on relativity as arbitrary and fictitious. However, the concept of space folding to curtail (mostly cosmological) distances, cannot be assumed to take time with it if the time is not part of 4-D geometry, and the same thing as space as given in the Minkowski formula for equating space to time. It is the most alluring theory in all history; no wonder mathematicians have fallen in love with it. Unfortunately he had to conjure it with imaginary time coordinates and therefore completely false.

To recap, the fact that events can go forward indefinitely is mainly responsible for the myth about Past, Present and Future that has been developed into a philosophy. Our memory mechanism has something to do with this. The day and night system certainly casts considerable influence on all notions of past and future, and time travel is said to be 'a scientific possibility'. You can't blame the mischief-makers for turning Past, Present and Future into a field of serious study about the nature of time. In fact, our memory mechanism is partly responsible, and while nobody can convince anybody that Past, Present and Future do not exist, I am certain that they can be logically explained to exclude the religion and mysticism with which they are regarded by some people.

Human beings are strictly controlled by the events that happen to them. These events have antecedents and continuing consequences---in other words, they have 'pasts' and 'futures'. Some use their memory and imagination to extend the antecedents to no end, or to project the consequences to

---

want to stress what Planck said, namely if the premise of any idea is incapable of proof, it is sheer (irritating) humbug to introduce it into science and logical thought and then ask serious thinkers to disprove it. The worst example of this is the Minkowski formula for 4-D geometry. He had to rely on imaginary time coordinates. This makes his theory logically untenable, yet mathematicians believe in it, and are asking question about how it can be mathematically or logically refuted—complete waste of time and mental energy.

infinity and even argue that by time travel you could meet your grandparents before they were born. Such excesses have nothing to do with time or the logical and scientific explanation as to why and how we have time. What can be said confidently is that the logical and scientific explanations persuade us to rule out any divine influence on the thing we call time. Whether God exists or not we have created time out of repetitive motions that owe nothing to Him, and that is what this book is all about, for that is what we can prove either logically or scientifically. And it is important that the time we are discussing in this book is exactly the time we have been using on earth, including even the atomic time, for that too is based on regular motions, pulses or oscillations.

Given the analysis above, the problem regarding past, present and future can now be conclusively resolved. It seems to be part of the religious notion that time is marching through nature from the past to the present and the future to come. But the religions offered no evidence for this other than the dreams of certain people interpreted as 'Revelations'. We have now scientifically discovered that time is not marching through nature. It is, in fact, discrete. We don't have to look far for the evidence. There is only one year. We repeat it to make time continue. Otherwise one year is the end of one long unit of time that we have had to subdivide to get the other fractions for cultural use. This mean there is no longer a universal time marching from the past to the present and to the future. Rather the past is memory; the present is now; and the future is mere expectation.

PROCESS AND REALITY

What we call 'time' is merely the passage of time because the repetitive motions or cycles we count as the rates of time are passing. They can only show how much time is passing. We estimate "how much time" by the number of cycles, say the years. Two cycles means two years. Some thinkers believe that it means time does not exist. But the point is the system gives us rates of duration, of being, of waiting, or periods of the passage of time; and we have nothing else to employ to guide our actions on earth. So we call it 'time'. It merely means there is something we use as time even if its true nature is unknowable. Materially, it is easy to imagine that there is natural time known as the period of waiting which may be due to the fact that chemistry takes time, and all things are created through physical or organic chemistry. The processing may be chemical, but we will experience it as a period of waiting, or time. Gestation, for instance, involves waiting or the passage of time. But in fact the whole nine months is the period it takes the foetus to go through the chemistry of life.

I consider Professor A.N Whitehead's "Process and Reality" the most profound concept in metaphysics, or the study of nature generally, from science to magic, particularly magic. Let me illustrate this with what happens in chemistry and medicine that we regard as the catalyst of time. Of course, medicine is based on chemistry; the whole human body is a chemical factory. And in chemistry time is known as a catalyst. In fact, the catalysis is the effect we experience during natural processing in the body. The processing is realised as 'time'. At the dentist, for instance, injection is normally given to numb an area of the mouth under examination so as to reduce pain. You are told to wait for it to take effect. On the surface, that is time. You are often told specifically that it will take about five minutes. That is time. But in reality it is processing, the processing of the chemical through the body---the process appears as the 'reality' of time. The drug is not sitting still in your body for the five minutes to elapse; it is in the process of active communication. It is going through you cell by cell, even down to molecules. That constitutes the waiting period, demonstrating the process and reality syndrome in nature. To the patient it is time. In nature, in the human body, as well as in metaphysics, it is chemical processing or a process---physical, chemical and mental. The process is realised as time, hence 'Process and Reality'. Whitehead was right. Process and reality can be used to explain the cause or essence of time in metaphysics, namely, what we experience as time (or waiting period) is, in fact, the processing in nature at the chemical, atomic or even the sub-atomic level of matter, the body, et al. In this short chapter, I merely want to point out that, to my way of thinking, this is the nearest we can logically get to the scientific definition of time. How long we wait for anything is just how long it takes to process something chemically.

I believe that due to the day & night system, time is a practical issue. At the same time, due to its closeness to ageing it can also be seen as a metaphysical one. But, thankfully, through the rise of science (especially of chemistry which is the basis of physics), to annoy those philosophers who still cling to the Mill type of bypassing science completely, some of us can now see that time is less of a metaphysical mystery. We imagine that chemical processes give rise to the period of waiting we experience as time: in nature it is chemistry (physical or organic) going on below the surface without human intervention, even without human awareness; but in our senses it is a period of waiting called 'time', which, when quantified as discussed in one of the essays in this book, can be mechanised as time in the clock ticking away as discrete time in specific units. This shows that there are two aspects of time. On is the psychological (subject to emotions, religion, myths, ancient legends and so forth); the other is the physical aspect, or the physics of time, which are the cycles we count as the passage of time and mechanise into the clock. Post-relativity time (which means secular time or the notion that time originates on this planet), as I am trying to show, must attempt to provide a link between the mental and the physical aspects of time that leaves nothing to be explained with reference to God.

8

TIME AND QUANTIFIED TIME--- OR THE PASSAGE OF TIME.

We know time in units; also it can only be used in units.[123] *We know it in units because the quantification method (being repetitive cycles or motions), reduces it to units---year after year after year; or in seconds minutes and so forth; for they are fractions of the year and therefore also part of the yearly cycle. What is known as silent time and therefore not delivered in units is chemistry, physical or organic; the best example is ageing. The religions stress this but they are wrong because when you come to tell 'how much time?' in all situations you need it in units; so creating the units beforehand (in all cases) is crucial in the study of time. This involves mathematics. It means we quantify time. My argument is that what we use to*

---

[123] Many of the ideas in the book have been repeated over and over again, because, although relativistic time is easy to explain in physical terms (e.g. orbits of the sun or atomic oscillations), it is necessary to hammer home the new ideas to replace people's inherited instincts and thousands of years of thinking that time is universal, fixed or absolute and providential. It is a thankless and difficult job, since secular time is definitely unpopular. People prefer to believe that the mystery of time provides hope that death might not be the end of life.

quantify time can only show how it is passing by. To put it differently, because we use repetitive cycles or motions, the method can only give discrete time. Discrete time is passing unit by unit continuously. Hence there is no need for any theory to account for the passage of time. The years need no theory to pass by---they are passing anyway! They are passing by in procession, as repetitions----second, second, second, minutes, hours and year after year after year--- always in procession to constitute the passage of the years as the process of ageing. This ageing is adding to our years, but the years are mere physical orbits of the sun; we count them as the rates of passing time because we have nothing else to use for time. Whether time exists or not, what we call time is just how it is passing by through physical cycles[124].

Let me begin by pointing out that we are all fond of using the word 'time' loosely to refer to the passage of existence in any form whatsoever. That may be called 'the unscientific notion of time'. In logic, science and philosophy, however, time is what Professor Richard Feynman called 'how long we wait'. This translates into the concept of 'how much time', or quantified time, so as to be able to tell how long we wait in mathematical language for universal application. The essence of my theory is that how we quantify time is the same as how time passes through nature, and since we know time from how it is quantified, all we can ever know of time is how it is passing through nature and never what it really is.

In a serious discussion of time, it does not make sense just to mention time. It makes time seem to be in the very air we breathe, or equivalent to existence, Being, motion or any passing moment. Either none of these is time or all of them constitute time---that, for me, is the conundrum. I agree that any passing moment or Being or existence is time but whose time is it, since we now know that cosmic time does not exist? Whose Being? Whose existence? Whose moment (that is, through whose perspectives), since we are so many and individualistic in our perspectives? So the context of any proposition (in science, mathematics and philosophy) must

---

[124] This section is a bit technical but it does not mean the discussion of time should be left to the Professors alone. Everybody uses time.

always show or imply the sense of 'how much time' in it, or expressly show the quantity of time proposed. And that is what I call 'quantified time' as I will presently explain.

Of course, time may pass when one is not conscious of it. But, in all cases, when one wants to know how much time has passed, or will pass (as in futuristic propositions), mathematics must be used to quantify the time. And let me stress again that we quantify time by the use of external cycles in union with any sense of duration of anything whatsoever. We have over the centuries acquired a sense of duration (being a psychological device in the mind for telling or marking the lengths of periods from one another)[125] so we know that one hour is longer than one minute. This is what I call 'quantifying time'---i.e. reducing time (whatever it is), to culturally useful units of duration in the mind (days, hours minutes, etc.) which are derived from the day and night system. It is what gives us expectations of 'coming time', that is, the long periods of the day and of the night that we know will always come from past experience. That is what we use, together with the coming year, to plan all activities in life.

First, what I call 'Quantified Time'. Quantified time is 'time in a clock', any clock at all---reducing time (whatever it is) to units of duration, as already mentioned. And the clock, any clock, can only show time as independent of space. Einstein is very important here, too. He initiated the idea; but it has been expanded. Logicians agree with the Lorentz local time notion, meaning that you get your time from your local space which is what Einstein said inspired his theory of time. (He said he took the Lorentz local time ideas as 'time, pure and simple'.)[126] While other scientists regarded the Lorentz discovery as 'time dilation', Einstein realised that it was the first time that time had been shown to begin from other places or situations---or seemed different to different people, as against the religious view that it was general and absolute, or fixed by God and the same everywhere.

---

[125] This instinct is certainly very old and not all that mysterious because the brain is very efficient and it has had thousands of years of practice. But, in ordinary human affairs, it adds to the mystery of time for the uneducated.

[126] We have to remember that, at the time, this was not only revolutionary, but a scandal or even sacrilege---as it was contradicting the Newtonian concept of absolute time.

This was a great philosophic insight. The important thing is that he called it "Time, pure and simple". *For it was not a mere dilation of an existing time; it was the creation of time or of a different sort of time from what was known as the normal time which was not supposed to changed under any circumstances.*

If it was a dilation then it was the dilation of one clocks, not the dilation of time as a whole. But as the first time anybody had shown that time can begin from anywhere, it was momentous. It changed our concept of time completely, making it human or secular. So he noted that, "There are as many times as there are inertial bodies".[127] Logically it means there can be as many times as there are (many) 'bodies'. The great logician, Bertrand Russell, immediately inferred that time was [therefore] 'relation between points'---for that is how we get our units of time. The year, for instance, begins from one point to another, and all other subdivisions of the year are obtained with points. It means we use points to derive units of time from space, which can only result in a time system that is basically independent of space. Minkowski could only change this with mathematics but not in physical reality. So we have to reject his formula completely.

The other aspect of time is its passage; but it is the same as the actual process of quantification--- that the process of having time (or what we call 'time'), is how it is passing by or how it passes by only and not what it really is. To illustrate this, one example that I am fond of goes like this: first, I insist that we have to get accustomed to a new definition of time as the rates of passing time and never what real time is; and the most logical illustration (to my mind), is the time for doing something, say, boiling an egg. It may be ten minutes which is a fraction of the earth cycle and therefore part of a physical cycle. Or it may be 100,000 units of rapidly passing atomic oscillations, not much different from tapping the finger. Yet they are 'time' for boiling an egg. What real time is we do not know; the nearest to that knowledge is how it is passing by, and so many units of repetitive cycles or motions amount to the 'time' for boiling an egg. As I have said, in my view, this is the most logical manner

---

[127] Abraham Pais, "Subtle is the Lord..." Opp. Cit. Ch.7. Most of these terms have been quoted elsewhere in the book and appropriate references have been cited already. I regret the repetitions but the reader must allow that this is a pretty difficult subject---the worst in the world!

of defining time as opposed to the many illogical ways people define it. The time for doing something is just a number of physical cycles---and they are passing. They are not static; and they happen to be all that we can ever know of time---for boiling an egg. Similarly, the time for doing anything (any duration) is a certain number of physical cycles. It is also the proper definition of time overall.

Space-time, as the Einstein notion can also be called, is automatically quantified as it is derived from space with points, which is the only reason for calling it 'space-time', as far as I am concerned. I have to stress that it is important to realise that quantified time is necessarily discrete---because the year is determinate and we have to start another year at the end of each one. Because the year is determinate, all of its fractions obtained from it with points (as fractions thereof) are also discrete. When we are told in special relativity to create our own time because universal time does not exist, as Russell has remarked, it amounts to using space for the time we could create. There is nothing else we could use if the time is to be in units as our time has to be (or even all time systems have to be). But time obtained from space can only use two methods: as relation between points, or by means of repetitive cycles. Both methods lead to discrete time, which, at once, solves the oldest and most mysterious quandary about time.

For discrete time can only pass through the succession of the individual units. On this point, Leibniz was absolutely right when he said time is succession. What was lacking in his day was the concept of discrete time; with this new concept in our post-relativity world, we can now see clearly as to how time passes and seems continuous through the succession of its separate and individual units: second, second, second; or minute after minute after minute. Plus the hours, weeks and months all the way to the year, which also passes in the form of year after year after year.

It may seem surprising, the springs of a thousand legends, giving rise to supernatural speculations, that we have an extremely ingenuously smooth time system, so cleverly structured that it is there when we are born and there as we die, and always passing by. For this reason we know that "Time does not wait for anybody", not even Kings and Queens and Presidents. Even surrendering one's Kingdom and all possessions for a moment of time cannot save the most powerful Queen on earth. Scrutinised under a logical gaze, however, time is not so rosy; it is only one moment, repeated to pass by and seem continuous so that arithmetic can be applied to its accumulations.[128] This, as we know well, happens when we reckon time for futuristic planning, and backwards as history.

But for the union between the sense of duration and external cycles giving us units of time out of the moments of time, time for the clock would not exist at all. Presently philosophers see time as rather a straightforward pragmatic entity, albeit not as simple as it is normally supposed. It is partly a confidence trick, which makes the clock work continuously, the trick of continuity is in the repetitions of the seconds, or of the units of time, all of which are to be understood as single moments---which are the realities--- of quantified time. It is also partly physical (using physical cycles for the process of quantification); and partly philosophical, or psychological, as Professor Whitehead put it, "A time system is a sequence of non-interacting moments."

---

[128] Let me explain that space-time is necessarily discrete. We have only recently come to understand space-time from Albert Einstein; yet time has always been discrete, consisting of only one unit (or moment) of time---of whatever length. For there is only one year, and all other units are obtained from the year in the form of separate units of time. To get more years we repeat the one year exactly. Thus we have second, second, second; or minute, minute and hours and so forth. Each is a moment (or a unit) of time in its own right. The notion is best illustrated with mechanical devices: if you set a timer to regulate the working of any mechanical device, when the time ends, the machine will stop because the time allowed has ended.

8(i)

WHAT IS MEASURED BY A CLOCK

In his book ABC of Relativity, Bertrand Russell asked this important question: if cosmic time is abandoned, what is measured by a clock? This is because he had already shown that under relativity there is no longer a universal time.[129] *But we do not measure time. We create it in units with repetitive cycles and mechanise the contraption in a clock for reproduction (so carefully planned and mathematically accurate) that the clock ticks about 31, 536,000 to equal one year, and start again for another year.*[130] *The ticking of a clock is timed precisely to coincide with the motions of the earth round the sun over and over again---that is not measuring time; it is counting units of time deliberated created so that a certain number will match the motions of the earth. The motions or orbits of the sun are not time. They are physical motions we count as the rates of time's passage. I repeat that human*

---

[129] Some writers (including Professor Yourgrau and his hero Kurt Godel) have seized on this to argue that if there is no universal time then there simply is no time at all, being unable to imagine how man could have created time. The point is, there is something we use as time. That is what I am trying to explain in this book.

[130] This was a major clue to the nature of time yet we missed it for centuries.

beings can never know the real nature of time.[131] *Everything we have ever use to reckon time has been a mere contraption 'to construct' time for us---as Russell supposed---without knowing that that was all our methods were giving us. That has always been the passage of time only. Poor souls. Until Einstein we were sure that we were actually tracking real time---not at all. We only ever got to know how it was passing through nature.*

Of the four notions of time mentioned above (the religious creation at noon on AD 4004---call it what you will—the Newtonian absolute time, the Minkowski space-time continuum and relativistic time of Einstein) only the last one from Einstein can stand the test of logic, for it is the only one that can be mechanised into the clock. Our time is based on the repetitive orbits of the sun by the earth, and evidently the earth never stands still.[132] If ever it does stop going round the sun, our time system will be completely nullified; but, of course, life will go on. It is inconceivable that all life will be extinguished instantly the moment our time is (mathematically) nullified in the sense that only quantified time would be lost. This is the best proof there is that life is not based on "time allowed", as the religions believe; rather time is a union between the sense of duration and external cycles---therefore man had something to do with the time we have in the clock, the only reliable time, as quantified time. We can therefore see that only the Einstein notion of time accords with our only means of gaining time---i.e. by regular motions.

All the religions speak of "time allowed" for the duration of a man's life. They had to, because the nature of time is easier to explain as a providential bounty than anything else. To be honest, without a cosmic explanation for time, what is time, or,

---

[131] It appears that this ties in perfectly with everything else in nature---that is, the real nature of things is never known. We deal only with their effects on our senses. So the nature of time is unknown---we deal only with repetitive cycles as the rates of the passage of it. On this point, Plato's 'Simile of the cave' was prophetic.

[132] I have always to remind the reader that atomic time is based on the same principle, namely regular motions or cycles to generate units of time that we count as the rates of passing time.

to put the question in another form, what is the origins and essential nature of time? Of course, it is assumed that the clock measures time---but from where? And what is it that the clock measures, or which we know and use as time?[133] The clock maker will say he invented the clock to reckon time in the sense that everybody knows---but what is that sense of time? Everybody assumes that time is just there. In fact that is not so. Existence, of course, is there; we exist. And time is closely associated with existence; yet since time requires points we have to realise that it is not the same thing as existence and try to establish its origins. This is easily proved: when the earth ceases to be capable of supporting life, who will be there to count the orbits of the sun as 'years'? And yet we know that life will end with the demise of the sun. Existence and time are not the same thing---that is, a unitary entity, because somebody must be there to set the points for the years, etc.

Time is deceptive due to the ticking of the clock. It makes time eternal. It is there when we are born, even before we are born; and it is there as we die. More importantly, all the philosophers have been deceived into thinking that the clock measures time; but that makes time universal: an entity existing naturally in nature of which we measure our own version.

In spite of the Einstein notion of time (that there is no longer a universal time and that every 'body' has to have its own time, and therefore there are as many times as there are inertial bodies), all science is dominated by the idea that we measure time, implying that universal time does exist, which is one of the reasons they accepted the Minkowski 4-D geometry and went on to invent something called "Time Zero". Yet a universal time

---

[133] Without the explanation that what the clock measures are cycles of duration, or duration reduced (or converted) to repetitive cycles to equip us with manageable units of time, metaphysically interpreted as a union between the sense of duration and its conversion to external cycles, time could never be logically accounted for. We would just go on using it---but in what form? Perhaps in the form of units (year after year after year, and all the seconds and so forth derived from the year); yet that means the same thing, namely, a union between duration and its conversion to external cycles. For the year is only an orbit of the sun. It is not time. It is the practice of humankind to call it 'a year; and we use it as our basic unit of time, as a matter of convenience. Otherwise, in nature it is not time. As a matter of fact, we can use something else---we can tap the finger, for instance.

is in violation of the Einstein theory of frames. In fact, the seconds are obtained elsewhere (through the logical analysis, or subdivisions, of the year as fractions thereof), and deliberately programmed into the clock for reproduction, so that a precise mathematical number of seconds add up to one year. I count the number as about 31,536,000---the precise number is not as important as the principle involved. In practice, mathematicians have cleverly designed the clock in such a manner that we can count the number of seconds for one year either forward or backward. Forwards for planning purposes, say, so many seconds adding up to minutes, hours and so forth will be required for an event or action of one kind or another. And backwards, saying something like: it was so many seconds or minutes or hours, days and moons ago that so-and-so happened. This is the ultimate logical analysis of time by the clock. Yet from the point of view of ordinary people the clock is seen differently (without logical analysis), thus spawning numerous myths, legends and even scientific theories, many of which are compounded by religious beliefs.

When it is postulated that general time permeating the whole cosmos (and therefore the same everywhere) does not exist, the first implication is that every material body or planet has to have its own time; it is not coming from the cosmos therefore it must have originated on this planet.[134] So let's find out how it all began. That is the first implication. The second is that, as a result, cosmic time is abolished---although it sounds tautological, it still has to be emphasised, as well, and most

---

[134] Let me state the nature of traditional time: that it is fixed by nature or God; that it is the same everywhere; that it is marching through nature since 'Creation' or evolution (scientists also believed in traditional time); that it had a definite beginning; and that it will end at a certain date in the distant future. Under relativistic or the Einstein notion of time, none of this is true of time. Those of us foolhardy enough to try and offer any sort of logical interpretation have to face two obstacles. One is the natural resistance of humankind (including scientists) to new ideas that unsettle their traditional views of anything. The second is to offer satisfactory, rational explanation of the origins of this new notion of time that is supposed to be limited to the earth---and why. But life is made infinitely worse for us the interpreters because Minkowski intervened with another new theory of time much adored by mathematicians and scientists; and although it is based on imaginary time coordinates and therefore logically flawed, they have fallen in love with it.

clearly because the 'cosmic time instinct' is permanently ingrained in the human mind. One reason is that time cannot be suspended; but the more cogent reason is sheer intellectual incompetence plus fear of the unknown. We are always using it, and so it does not make sense to just insist that it is not there. If it is there, and did not come from the cosmos, how did it begin? And the obvious fact is that it is always there. Even before we are born, and also as we die to leave it behind. Yet it cannot be supposed that each body's time is a version of something 'naturally existing', whether it permeates the whole cosmos or not, with the necessary but illogical (little 'academic') proviso that it may not be the same everywhere but varies with individual bodies in accordance with unknown natural laws since it is dynamic. What Einstein found is that space is dynamic and time is based on space. But the reality is that time in the clock is stable; and when the complications introduced by Minkowski is removed, we have to revert to time in the clock. The only reliable time is time in the clock.

Of course space is dynamic and time is based on space. But there are several types of spaces in the universe. The dynamic space in theoretical physics is not the space in our bedrooms, the space on the streets or the space in our skulls. Time is reckoned by means of the space we see in daylight not the dynamic space in theoretical physics. One of the mistakes of Herman Minkowski is that he regarded mankind as stupid; yet we are not so dumb to fail to distinguish between the several forms of spaces in the universe.

It is plainly evident that this erroneous sense of time dominates scientific thought. Hence, time is not defined in physics; and as a result, the Minkowski fiction makes sense to some scientists, including even Albert Einstein himself.[135] So far, only Professor Arthur Eddington has redeemed physics by warning that it must never be forgotten that the Minkowski formula is "fictitious and arbitrary"---but they have chosen to ignore him.

Thus, Russell's query is important, namely, "If cosmic time is abandoned, what is really measured by a clock...?"[136] My answer, of course, is that outside the union between the sense

---

[135] We are always reminded that he used it in his field equations of general relativity. Yet we know that he never accepted it as logically valid.
[136] ABC of Relativity, Ch. 4.

of duration and its conversion to external cycles, time does not exist to be measured.[137] The very act of 'measuring' is the time in essence---like moving from point to another point, time is going, so that time becomes 'relation between points'. The cycles constitute the time units (the years, for instance), and the time units constitute the time: a year is a cycle, but it is our time, the basic unit out of which all other units are derived.

However, the cycles are the creation of man for the sole purpose of converting the sense of duration (of anything or any event, like the period it will take to reach the village from the farm before nightfall to avoid predators), to time units to guide his activities. So the clock does not measure time; it rather reproduces units of time programmed into it repetitively---second, second, second, and so forth. It should be remembered that the seconds are put there by the clockmaker; but where do they come from? The answer is that they come from the subdivisions of the year. Otherwise the time does not exist anywhere to be measured---the units constitute the time. Without the year there will be no seconds, and the like, all of which are derived as subdivisions of the year. As hinted above, you can even dispense with the year and its subdivisions and tap your finger, if you will not get tired. A million taps means it is time to go to bed, and so forth; outside the units of time, time does not exist to be measured; but the units are the creations of man as quantified time to record (trace or track) the passage of time.

In conclusion, let me try and make the argument a little clearer with a question: where do the units of time come from? One answer is that they are obtained as fractions of the earth-year in association with some of the features of the globe---day and night, the moon phases, and the seasons. But where does the year itself come from as a unit of time? In fact, it is only a physical orbit of the sun, but we count the years as the rates of passing time, hence the centuries. We have nothing else to use for time, and even this took us ages to perfect for reckoning

---

[137] I agree with the supposition that time does not seem to exist physically. My theory is that it cannot be known physically but we have created something we call and use as 'time' with regular cycles thus making that something we call time automatically quantified and discrete. Because of the Russell query about what is measured by the clock, I regard this section of the book as the most important.

time. Otherwise there is no time in nature---sentience is required for counting the earth's physical orbits of the sun as years, and pare the year down to the seconds. By quantifying time we get figures to simplify it for all and sundry hence the clock. The problem is that without quantification we couldn't get time in units to create the clock; but quantification makes it discrete (which is momentous as discussed in this book); it also makes it essentially secular and also passing non-stopped, so that nobody can ever know what its real nature is other than how it passes through nature---like the passing years, together with its fractions and its replications to become centuries.

Furthermore, for purposes of clarity, let me give another example of how what we know as time is merely our process of quantifying natural time for our cultural use as discussed in the chapter above. It may be asked whether since time is changeable (for that is what Lorentz discovered),[138] can we change 100 year to 10 years numerically, maybe for people to prolong their days on earth? The answer is that we can do that by mathematics; it's just a matter of changing the figures. But is it so in physical reality?

We can change the figures, but what of the chemistry of things during the passage of those hundred years? This goes to prove that there is natural time and it is probably chemical in nature. For we can change the 100 years to 10 in mathematics, but we can never change the body of a 100 year-old-person to that of a ten-year-old child. The chemistry will have moved on. The implication is that life moves on; that is what we call ageing---but it is also time moving on naturally.

To keep track of this natural time we can only use repetitive cycles or motions and count them as the rate of the passage of time---the years, for instance, as pared down to the seconds, etc. Thus the passage of time is known to us only through the quantification of natural time---so the passage and quantification mean the same thing, and there is no need for

---

[138] The discovery led to the Einstein notion of time to the effect that it is neither general nor absolute. All interpretations of time since then, to be rational, have to take note of this fact. Academic philosophers are not aware of this because they delight in keeping themselves ignorant of relativity or use the Mill method of bypassing scientific activity---a closet religious objection to science, especially Darwin.

the arrow or arrows of time idea. The process of quantifying time makes it necessarily discrete---year after year, etc. Discrete time needs no arrows to move it on. It simply replicate, that is the reason we have our time in units, and in units only. For time does not exist for us except that we have created something we call 'time' through the use of repetitive cycles, in which case the units become "relations between points", as Bertrand Russell observed.[139] The year is from one point to another. And there are no years, only the one year replicates all the way to the centuries and beyond. So it is not true that there is no time, as Professor Yourgrau argued in his book under reference.[140] The truth is that we have time but cannot define it unless we use regular cycles and count them as the rate of the passage of time---but the rates are physical units of space, just like tapping the finger. It is we who call them time, because we have nothing else to use, and employ them as guide to action: Ten physical cycles means it is time to do so-and-so, as we use the minutes and hours for because they are also cycles as fractions of the big earth circle. All the time natural time is passing so that we age.

Frankly the academic world has not served us well on the subject of time. It is not surprising when CUP could say they do not serve the general audience in philosophy. They serve only the academic philosophers, like Wittgenstein, whom the university appointed professor even though, according to Bertrand Russell, he was preaching nothing but logical mysticism that puts an end to physics. Can any honest and competent academic thinker accept that as worthwhile in a professor of philosophy? Oxford replied with William Kneal who advised us to "retain the Platonic notion of Mental Events which are distinct from anything in the physical world..." Thus the two

---

[139] This is a new idea. There is only one method for promoting new ideas, namely by repetition for somebody to take pity for the perseverance.
Normally the repetitions are buried in academic journals, but the academics ignore me because of my attacks on Minkowski whom they adore.

[140] The orbit of the sun is a physical activity. We take it as one year and pare it down to the seconds and use the fractions to guide all human action on earth otherwise we couldn't live. But it is just like tapping the finger and counting them as the rates of the passage of time; thus even the atomic oscillations are also the same sort of time keeping. They are all we know as time; but it does not mean time does not exist, only that we can never know what it is other than to guess that it is close to chemical processing.

'greatest' British universities were competing with each other in mysticism: one was the Wittgenstein 'logical mysticism', the other was the familiar Platonic mysticism already refuted by G.E. Moore---who was replaced by Wittgenstein. Without the Cavendish laboratory Cambridge is medieval in attitudes; and without the OED science and medicine is an aristocratic, political institution with no merit and little contribution to rational thought.[141] But then we have to remember that, until the rise of science, and particularly of Newton and Einstein, these two so-called 'great' universities were mostly extensions of the faiths. At all times mysticism is soothing for after all you make it up yourself for your own comfort---e.g. no religion has a cruel God, but always a kindly saviour to be worshiped for salvation.

My point is this, once it was discovered by Lorentz and Einstein that cosmic time does not exist, and Bertrand Russell asked the most intelligent question in human history ("If cosmic time is abandoned, what really is measured by the clock?"), it was time to investigate what it is that we know and call and use as time on earth. Some were saying it does not exist at all; others were claiming that time travel is 'a scientific possibility'. But nobody was working out a coherent logical theory of the reason we have something we call and use as time and without which civilization could not be in existence.

This defective book is my own fallible attempt at a solution, except that I want everybody to note that what we call time comes from the use of repetitive or regular cycles---like the year for instance. This is not a new idea; for that is precisely how Professor Richard Feynman also described time about twenty years before my time. I further remind the reader that there are no years in nature. We simple repeat the orbit of the sun over and over again to get all the centuries. This means the basic unit of our time is (a) obtained as a physical unit; and (b), strictly determinate. We find also that all other units of time

---

[141] Of course, they do have enormous funds and do carry out some good work in science, engineering, medicine, even in astronomy and mathematics since they attract some of the best intellects in the world; but in other important subjects that determine the roles of human beings, of nature and of what reality is for the rigorous rational study of the world, these institutions remain rooted in the past and the faiths: one, the philosopher is urging us to follow Plato and his mysticism and its religious connotations; the other, CUP, printed it in a cheap paperback edition for mass circulation.

are based on or related to the year as fractions thereof, and being derived with mathematics or points, they are also in units (seconds, minutes, hours, days, months and so forth.) We put all this together and get discrete time---created with points. And for time to pass by discrete time can only pass through the procession of its units. The rest is inferences and interpretations, legends and things we call time in the light of this theory which is pragmatic and not at all as metaphysical as other weird interpretations of time. Pragmatic because the Lorentz/Einstein notion of time came out of their experiments as empirical ideas not metaphysics. The discrete nature of time is also the way we get the year and repeat it over and over again. Yet when I put this to the academic publishers[142] they turn away mumbling something about Minkowski. But Eddington, Whitehead, Russell and even (we are now told) Einstein never accepted that the Minkowski formula for equating space to time is logically valid. Of course the subject is difficult; but academics are supposed to hammer at a subject till we understand them; they are allowed to spend millions of our money and years to find things out for us. However, in the case of time they do nothing of the kind. They rely completely on the Minkowski arbitrary theory---is that not a disservice while they chase media money? Of course, at my age, I am too old to fear mockery, criticisms and ruinous sanctions--- otherwise I wouldn't dare!

---

[142] I have piles of letters mostly referring me to other publishers because they do not have the necessary expertise!

9

THE PRINCIPLE OF MATHEMATICAL EQUIVALENCE

Mathematics can only reveal what is there in nature. It cannot create something new in nature. It can reveal something new but only upon the basis of conceptualizations. The concepts come first. Mathematicians ought to respect concepts and set the successful ones in mathematics. The excessive claims of mathematicians are leading physics astray. Setting what is not physically accurate in mathematics as a mere exercise of mathematical acumen is nothing commendable. But it is true that mathematics is man's most efficient method for the study of nature's secrets.

In nature, there is reality and our perception of it. In the word 'perception' everything man does in life is implied, including mathematics, since we can only act by perceiving the true nature of the physical world; I am using the word in a sense akin to 'experience'. The problem is, pure mathematicians normally are permitted to imagine things to satisfy their nostrums, so that they do not rely solely on their percept; however outrageous, they can defy reality, logic and common sense, and leave it to the applied mathematicians (physicists, astronomers and cosmologists) to find out whether what they have assumed is there in nature so that their theories based on them can be seen as true or not.

In no other profession is this sort of thing allowed. As noted above, even one of the greatest mathematicians Britain has ever produced, Professor Sir Arthur Eddington criticised this common mathematical tendency. I have quoted him above in the text, but it will do no harm to repeat it as it is vitally relevant here. He said: "The pure mathematician deals with ideal quantities defined as having the properties which he deliberately assigns to them. But in an experimental science we have to discover properties not to assign them..." The principle of mathematical equivalence should make them think of the practical consequences of their imaginary properties, although I doubt it, but that is another matter. The rule is that mathematics should not seek to make the basic features of nature what they are not quantitatively; any such propositions are bound to falter. Note that we are talking only of basic phenomena. By the very nature of man, it seems he can make qualitative changes in peripheral nature not quantitative changes in the fundamental aspects of life, and time is the second most fundamental feature of life.

The principle means, in effect, one cannot use mathematics (sometimes defying comprehension) to state, say, that there are ten trees in a field, and propound theories about them if, in actual fact, there are only two. This is slightly different from assigning imaginary properties to nature. It is different because it relates to 'quantities'. If the supposition is only an assumption, it should not be disguise as a fact with bemusing mathematics. The principle of mathematical equivalence rules that, to accord with physical reality, one can only talk about two trees, or as things are, not what the mathematicians want them to be. Nature is not there for the convenience of mathematicians; it is neutral. That was the advantage we gained when the ancient teleological interpretations of phenomena was discredited.

In order that what I am saying cannot be misunderstood, let me make plain that we all know the importance of mathematics. Of course it is not only useful but vital for human existence. Without mathematics hidden structures and dangers, material compatibility and connections, strengths and weaknesses or defects, cannot be established. The computer and videos, the

TV, power generation, air traffic control and air travel together with thousands of mechanical aids to human life on this planet could never have been possible; or they could not come to exist at all without mathematics.[143]

Our war efforts during the last ruinous world word could not have been so successful without the breaking of the German codes by our clever and hard-working mathematicians. Navigation on land, sea and air, GPS and the manufacture of millions of things, from ordinary sheets of paper to household gadgets---all of these use mathematics. The ordinary door nail would take ages to produce in any quantity if we have to measure each one without mathematical aids. I have stated above that our time units have been obtained with mathematics as fractions of the earth-year, and that is true. Otherwise we couldn't have units of time for cultural use. Mathematics dominates modern industrial processes so completely that even objects that are agricultural depend on mathematics for weather forecasting or irrigation projects. It does not, however, mean the world or nature is mathematical---concepts come first and are more important than their mathematical expressions. Concepts cure our ailments, concepts entertain us; they inform us; they educate us; they create love and hate, war and peace, all social contacts.

So we can all agree there are countless activities we could never have contemplated without mathematics. But mathematics and conceptualization are two different mental activities, and I think the concepts come first. Let us take one small example. Suppose you have to share ten pineapples among one hundred candidates. By mathematics you know that the ten pineapples have to be divided into a hundred segments, meaning that each has to be cut into ten parts, simple. But mathematics will not tell you why, that is, the medical, chemical or nutritional reasons the people involved have to eat pineapples. The reasons are concepts that we derive from other branches of human knowledge where logic and the meanings of words are paramount.

---

[143] Of course, efficient weapons of war and weapons of mass destruction could not be invented either. However, the use of science for wars is a problem for the politicians not scientists.

Again, very often we hear that philosophers engage in arguments for argument sake and contribute nothing to progress, understood as 'scientific progress'. This is because it is not often realised how progressive is the study of philosophy; or how educative, informative and 'scientific' it has become since Bertrand Russell. Quietly but surely, with the help of philosophers many entrenched myths from our primitive past are being discredited one by one. Prominent among these is teleological argument. With that and many other ludicrous intellectual fashions out of the way, it is unacceptable to regard any concept as 'compounded for the convenience of the mathematician', as Russell accurately described the Minkowski theory of space-time. Someday, we may get scholars writing about the many myths philosophers have discredited through their quiet researches to foster science and progress generally.

So I regard this principle of mathematical equivalence as a strict and necessary doctrine to prevent mathematicians arrogating the power and right to alter nature quantitatively in the fundamentals of physical reality. We shall, and should, continue to alter nature qualitatively to our benefit---gardens, buildings, roads, cities, waterways, canals, railways, bridges, tunnels, all science (bar destructive devices), and all art, sports and so forth. They do not change nature but beautify it; but quantitatively, never. We cannot make one object two, or two objects one, physically. It is not possible. Not in reality only in the imagination or in somebody's mathematics.[144] The only one I know of that has achieved any kind of academic fanaticism, oddly enough, in the strictly rational post-relativity era, is the Minkowski formula, but then it is regarded as fictitious. Thus mathematicians who rely on it must know that they are falsifying their nostrums.

The only advantage imaginable in the Minkowski formula is the replacement of the 3+1 formula with the 'ict' equation. It is true that the 3+1 system takes time as an external entity and adds it to the other three natural dimensions of space or matter; while the equation of space to time in the ict equation incorporates time in space as the natural order of things. But whereas adding time to matter in the 3+1 system may be seen as an 'act

---

[144] They call the Minkowski mathematics 'beautiful'. Perhaps it is but false in so far as the outside world is concerned.

of interfering with reality' (especially when we can't even define time or know where it comes from in the absence of universal time), using imaginary time coordinates as an attempt to equate space to time is equally illogical. In the other formula at least we know we have something that we use as time, whatever may be the provenance. There is time. We use it daily in sports, work and travel and even for sleep. Since we use it we should know how to invoke it; and there is only one logical method for getting our time, namely, as everybody can see plainly, by regular cycles---such as we get the year. But employing an imaginary element in the determination of the nature of physical reality so that it appears on paper to be four-dimensional (when it is really not so in nature), is a more serious default in our suppositions than the defect in the 3+1 formula. We can honestly admit in the latter that it is the best we can do with the limited brain power we have; while the former situation would appear to be trying to fool nature---much more difficult than robbing Fort Knox in broad daylight.

What beats my understanding is the pretence (and they know it is a pretence because it is not logically correct), of scientist that there is such a thing called "space-time continuum", so that they refer to every space as 'space-time'. The experts at CERN announced on television that they were going to review the concept because they thought that c had been breached.[145] One can only hope that they'll keep their word, because there is a major problem with time. It cannot exist in the metric of general relativity or in interstellar space. The notion of time in everybody's head is earth time. Atomic time is based on the same principle of using regular cycles to tell the passage of time as we do in the case of earth time. The very idea of time that is assumed to be everywhere 'since the beginning of time' is logically invalid.[146] Whose time is that since according to the

---

[145] They have hurriedly made another announcement that c has not been breached and that there was an error in their experiments. My point is that if the Minkowski theory is wrong (don't take it from me) it is so for the reasons stated by Russell and Eddington a hundred years ago---that it is wrong because it is based on imaginary time coordinates not because of the c in his equation. It is wrong no matter how light or any other entity behaves. These anomalies have been predicted by me for years; and I have always blamed them on the Minkowski fiction because of the i in his equation not c.

[146] The statement should be 'The Beginning of Existence'. For the two statements are not identical. The beginning of time on this earth came after

Einstein theory of frames every 'body' has to have its own time? We can and do use mathematics to adjust earth time and make it applicable to other metrics---but that is wrong.[147] Every 'body' has to have its own time that accords with its peculiar parameters. Many of the difficulties experienced in physics may be caused by this mistake about time. If the term 'space-time' is what scientists adore, they can keep it but only in the sense that time is derived from space, not that it is the same thing as space.

The origin of the rule outlined in this chapter will help the reader to understand it well when spelt out: it occurred to me when I was pondering Hermann Minkowski's claim to have made time and space into one entity as from the moment he outlined his theory, as previously mentioned, in the cheeky (even outrageous) statement: "The views of space and time which I wish to lay before you have sprung from the soil of experimental physics, and therein lies their strength. They are radical. Henceforth [that is, from the moment of his lecture] space by itself, and time by itself, are doomed to fade away into mere shadows, and only a kind of union of the two will preserve an independent reality". This is to make two things in nature into one with mathematics ('a kind of union of the two...') so he knew they were two independent things. How could he have made them one from the very moment of his lecture? He spoke of experimental physics. In fact, the only experimental evidence pointed to time being 'local' in nature; and Einstein adopted it in his special relativity.

There was no suggestion that time had been found to be inextricably intertwined with space---rather the suggestion was that time could not be had without space; and that once you

---

we learnt to count the orbits of the sun as years, not before. For after all, we were apes without knowledge of time for many centuries. But long before that we had been in existence. Similarly, the cosmos, however it began, is not the beginning of time because there is no longer a universal time that runs through the entire cosmos. What we call time is our own device ('construction', in Russell's phrase) for telling the time and will disappear when the earth ceases to be capable of supporting life.

[147] It may work in practice because time that is based on the orbits of the sun (as our time is) has not changed. However it should not be used for the interpretation of reality by saying, due to the Minkowski mathematics, space is four-dimensional, simply because physically it is not.

have space, you can create your own local time. What Einstein did was to interpret local time to mean "The only Time" we can have. Alternatively, every time is somebody's local time as the time for his inertial frame---because "there are as many times as there are inertial frames." People were astonished by all this. In the past they were sure of time running all through the universe of which we draw our version with our mathematics. It gave people the hope that death might not be the end because nobody could define time or how it came about other than the religious view which involves God and 'Creation'. They were surprised that a mere human being was telling them that this cosmic time does not exist, and that every inertial body has got to create its own time. Those of us trying to find logical interpretations of this new theory of time may even be annoying people---they seem to want to go on believing in universal time; and that is the reason they adore the Minkowski formula so much, particularly the closet religious scientists and mathematicians. One irrelevant question is that if Minkowski's theory is wrong why does his time works for us? The answer is that what we were using to track time has not been replaced; we are still relying on the earth's orbits of the sun for the years. What has changed is the insight that this is all the time there has ever been and that its logical analysis shows that it is all the time we can have---namely calling physical cycles as time, when what we mean (or what we are getting) is that they show how time is passing by only and never the true nature of time.

Going back to Minkowski, the actual physical reality known to be in existence was precisely as he stated it--- that time and space were two separate things. However, it is interesting that he sought refuge in experimental physics. In that sense, he did not breach the principle of mathematical equivalence. It shows that he was really a very good thinker; he had to be that good to convince Einstein to adopt his formula for general relativity, which came ten years later, even though logically the formula is not valid, and all the signs are that Einstein knew it. He knew it; but at the time he was facing pressure from mathematicians, who, presumably, convinced him that saying time is inherent in space made relativity easy to understand.

We must always remember that the 3+1 formula means man is the one to add the time---but, after Einstein's notion of time, the question is, whose time, or which time in the absence of a universal time? Minkowski answered the question with his fictitious formula equating space to time automatically. Yet if the great British scientist, Eddington, could judge that it was arbitrary (even if we can forget about Bertrand Russell's reactions), then we can also safely assume that all 'good' mathematicians (at that level) knew as well that the theory was logically flawed and still supported Minkowski--- probably, I believe, for religious reasons.

The unfortunate thing for Minkowski is that the evidence he cited was irrelevant to the claim he was making. He needed physical support that time and space are inextricably intertwined and therefore constitute one entity. The evidence that had been discovered by Lorentz and Einstein was that time was essentially local in nature, leading to the supposition that 'there are as many times as there are bodies', or localities, and that, additionally, by implication, time is different in different places, and also under different conditions. The principle of mathematical equivalence can be used to refute Minkowski's claim to have made them into one entity as from the moment of his lecture.

The rule stipulates that he could only have spoken about time and space as they actually were in physical reality, which, he has admitted, were two separate entities. The reality before Minkowski was that there was space, and there was time. Even the great Einstein himself made them independent in his special theory of relativity. It did not surprise me that Professor Sir Arthur Eddington and Bertrand Russell described the proposal as arbitrary and fictitious. However, it did surprise me that mathematicians ignored this strong condemnation to claim that they could not understand Einstein's ideas without the Minkowski theory.

This made me sit up and think, think of a principle to require mathematicians to relate their suppositions to exactly the nature of physical reality laid out before them, not as they would wish it to be to accord with their nostrums. The result is that I came to the conclusion that mathematics can only mirror reality, not to alter it with mathematics alone. That mathematics can only mirror reality was already known; I am not claiming any originality here. I am merely pointing out that after my analysis of the issues involved, I also found that Minkowski could not have changed physical reality to one of 4-D geometry with mathematics alone. Secondly, his mathematics was even defective. To get out of trouble, I prefer to use Einstein's own words for describing it. He wrote: "...The world is in this [Minkowski] sense also a continuum, for to every event there are as many 'neighbouring' events (realised or at least thinkable) as we care to choose..."[148] Einstein was joking, of course, even though he praised Minkowski in the same section of his book. Yet he knew (must have known, it couldn't be otherwise), that you cannot rely on the act (or vague notion) of thinkability as the technical reason for saying space is equated to time.

Therefore, the principle of mathematical equivalence is this: Mathematical statements (or equations) must strictly accord with physical reality. It means no mathematical quantity can exceed or reduce what the actual physical quantity is. No mathematics can make one thing two, or two things one, without physical divisions and unions. Applied to Minkowski, he failed because, as Professor A.N. Whitehead has pointed out, time and space still pass through nature as two entities, not one.[149] He could not achieve the physical union---it was, alas, only imaginary.

---

[148] Albert Einstein in RELATIVITY, Part One, Sect. 17 (Routledge Classics, 2001.)

[149] See his book, The Principle of Relativity, Cambridge, 1922, Ch, 1V

10

WHY SPACE ON ITS OWN IS NOT 'SPACE-TIME'

If Minkowski is wrong, then 4-D geometry or the four-dimensional space continuum (the Minkowski space or Minkowski universe) does not exist. And the Minkowski formula is described as arbitrary and fictitious. The alternative is to state that it exists in his mathematics only, and therefore should not be used for the interpretation of physical reality.[150] The fact that space is dynamic in relativity has nothing to do with the truth or falsehood of the Minkowski formula for equating space to time. His theory is arbitrary, meaning there is no physical evidence for it, no matter the nature of space.[151]

---

[150] Mathematicians know that the formula is false because it is arbitrary---meaning there is no physical evidence for it. In this case, therefore, insisting that it can only be refuted with mathematics is plainly disingenuous. The imaginary time coordinate upon which the theory is base does not exist simply because it is imaginary.

[151] The reality of time is that it has to be quantified with the use of repetitive cycles to get it in units. Theoretical space, we agree, is dynamic, but it is not involved. We are dealing with the space on the face of a clock. To bring in theoretical, dynamic space and add imaginary time coordinates to give us time in the clock is simply not feasible. It can be done in mathematics, but

On the nature and passage of time and 4-D geometry

In Einstein's special theory of relativity, we learn that, "In the absence of gravity, space and time are distinct entities. In the metric of special relativity they play distinctive roles."[152] We are still living in a special relativity metric, and nothing has changed in the Einstein theory since then to make all space "space-time". Yet in all their suppositions cosmologists and astronomers always refer to every space as space-time. I am deliberately stressing this over and over again because this book is specifically aimed at arousing the conscience of scientists about the logical anomaly of accepting the Minkowski fiction to distort relativity and physics as a whole.[153] I am launching a campaign! The CERN would not seek to question 4-D geometry as their first line of investigations when they encountered difficulties if they thought it was as sound as QED. So why are we stuck with it?

H.A. Lorentz (in his attempt to interpret the Michelson and Morley results), found that time runs slower when in motion, known as "The dilation of time as a measure of moving clocks". In a short digression, I would like to say something about physics and philosophy that occurred to me while pondering the Lorentz results. What he found is not that all time dilates at all; he merely noticed that to those looking at 'the moving clock' it seemed to run slowly; this has already been explained with mathematics; yet we are still told that time dilates as a measure of moving clocks. Those travelling with the moving clock noticed no difference in its performance; but to those observing it from outside the moving vehicle, it seemed to run slowly. To those biased scientists mesmerised by Minkowski that means time runs slowly with speed, and they have worked it into a

---

then what about time in the clock?

[152] Professor Jeremy Bernstein, in *ALBERT EINSTEIN: and The Frontiers of Physics*, Op. Cit. p110

[153] My original title for this book was "On the nature and passage of Time". I added 4-D geometry because publishers (mostly from the universities) replied to my submissions referring to "Your book about 4-D geometry..." At least they read me correctly! Mathematics exerts such a powerful influence on the human mind. If you are good in mathematics everybody assumes you are a person of super intellect. They are wrong. Those who conceive concepts are the brainiest---Einstein was not very good in mathematics, yet look at his intellectual achievements? I can employ a mathematician to rewrite all my books in mathematics---yes, but the ideas come first.

very complex theory of the nature of time, namely: time dilates with speed, and since time travel is 'a scientific possibility' by the Minkowski formula for equating space to time, the twin paradox is true, and so forth. Yet that one clock's performance (in the Lorentz experiment) is not the performance of all time per se. To a logician or philosopher, the people using the clock in the moving vehicle noticed no difference in its performance, therefore time is not involved; only one clock, being observed from outside, was behaving strangely due to the nature of light and the speed of the vehicle. So time does not dilate at all. Our units of time are fractions of the year that we have worked out with mathematical precision together with obvious astronomical features and therefore cannot dilate unless the year itself dilates, which means unless motions of the earth are altered.

Anyway, for the time being, Lorentz could not understand the reason for his experimental results: why this could happen in a world governed by absolute time, and literally put it aside. This is common in science. When something turns out too strange for words, you put it aside for the time being. He called it 'local time' or t^1. To him it was not 'the true time' but a mathematical auxiliary or curiosity---not very important. Time, he said, was time, denoted with t, and t^1 was something you get as your local time, but certainly not applicable in the outside world as time, because it was a mere mathematical curiosity. Lorentz later admitted that he failed to discover special relativity because he did not regard time dilation as of any importance.

Strangely, however, as one of his brain waves, Einstein worked this into his theory of frames. The dilated time was 'local time'---like somebody creating his own time---the time of his locality, because time can be created by anybody and from anywhere.

Now, if the universe is fragmented (with each fragment obeying its own natural laws as per special relativity), then local time would be somebody's time, which to him would be running normally like any other time, but to outsiders would be running erratically (or slowly, in this case.) In fact, that was precisely the case with the Lorentz discovery. People outside the moving clock saw it as running slowly; but those carrying it in the moving vehicle noticed no difference in its performance. That is the genesis of the Einstein theory of frames.[154] Otherwise, time

---

[154] Frames differ from one another so the parameters they use for their time

was separate from space.[155] What you will find is that it varies under different conditions, simply because everybody has to have his own 'local time' in his locality or inertial frame by his own peculiar parameters. But since time is continuous, and having made it a separate co-ordinate in the study of phenomena, in any one system dynamic space would have different time co-ordinates at every turn. We recall that Bertrand Russell has stated that from the sun's point of view the tram never repeats a former journey---because the time co-ordinates would be different.[156] Since time is a separate co-ordinate in the determination of physical reality, a different time co-ordinate implies a different situation, a different physical reality.

This was the state of affairs when Hermann Minkowski came in with his theory of 4-D geometry (or the space-time continuum), making time part and parcel of space---all space. It is easy to see why some people adore it. Soon cosmologists and astronomers call his theory "The Minkowski Universe", meaning that all nature is subject to the 4-D geometry, where time and space constitute one entity, leading to the abolition of the 3+1 formula and thus, in effect, reinstating universal time without saying so; and since the time system based on the earth-year remained the same it worked in the sense that theories, like general relativity, that used the Minkowski formula worked as well as those using the 3+1 formula---here was proof that time is already incorporated in space! Minkowski was hailed as the mathematical genius who made relativity accessible to the scientific community struggling with it. But let us swiftly add that the foremost mathematical interpreter of relativity was our own Professor Sir Arthur Eddington, the man who confirmed the general theory of relativity. He wrote the definitive book on relativity, called The Mathematical Theory of Relativity. About the Minkowski 4-D Geometry, he stated clearly on Page 9 (Ch. 1.1.), as already quoted, "Such a mesh-system is of great utility and convenience in describing phenomena, and we shall continue to employ it; but we must endeavour not to lose sight of its fictitious and arbitrary nature." He was not the only great

---

systems are bound to differ, and as they differ the time too will be different.
[155] The almighty conundrum is how we get the time or what we call 'time', if it does not come from the cosmos as a providential bounty?
[156] See Russell's Analysis of Matter, Ch. VI

mathematician who described the Minkowski formula as arbitrary. Bertrand Russell also said it was based on arbitrary assumption. As quoted in the book, he made it plain that because of that the derivation of the Minkowski 'interval' as time from space was illogical, or invalid.

Let me try and explain again the reason mathematicians still adore the Minkowski theory---even though they know it is fictitious. The reason is this: it makes things easy for them---created for the convenience of mathematicians, as Russell has observed. Yet it is not true. They accept the novel Einstein notion that time must be made a distinct co-ordinate in the description of phenomena. The problem is that, at the same time, Einstein made all time (any sort of time) 'local time'---the time you create for your own local purposes, as Lorentz had discovered. Einstein extended the Lorentz idea to all nature, hence his theory of frames. **With the universe being fragmented, it was impossible that one system of 'dynamic time' (as opposed to 'absolute time'), would apply with equal validity to all fragments of the universe.** As a result, he said there are as many times as there are inertial bodies in the universe. Nobody can contradict Einstein on this matter. However, mathematicians found that creating your own time to add to phenomena to acquire your concepts of physical reality puts too much power in the hands of man. (I suspect there are religious sentiments in this.)[157] Besides, it was complicated. The Minkowski system was easier;[158] you just have to mention the Minkowski space or ds^2 and move on. It comes with time already embedded in space as part of it---so the whole of space is 'space-time' and every time is also 'space-time'. The

---

[157] The Minkowski formula makes time universal again after Einstein, namely as (s=ct); something in general existence mysteriously (harking back to Pythagorean mysticism in mathematics), which can be invoked with the appropriate mathematical symbols; not as something you create in your own local space with the application of points to space, which makes time completely secular. It seems to me that humankind is not ready to accept time as purely secular. Those of us who have already made the necessary psychological adjustments for accepting time as plainly secular are not regarded as normal.

[158] It was difficult in mathematics but easy in logic and philosophy; and let me hurry to add that, because of the involvement of time, the whole notion of local time or space-time has philosophical implications, since time is the second most important thing in the world.

caveat of Professor Eddington was quietly ignored. Soon everybody forgot about this; Eddington and Bertrand Russell were dead; and there was nobody clever or bold enough to notice the discrepancy and question them about it. Of course, that leads to a distortion of relativity, but mathematicians are the arbiters of truth in mathematical physics and they were the ones benefiting from the Minkowski theory and therefore preserved it. In theoretical physics it is easy to cover things up with counterintuitive mathematics. Otherwise it is not true that all space is 'space-time', while all time is also 'space-time'.

Yet it is true that time is always space time.[159] You cannot have time without space; not because the space comes with time inside already but because all time is known and used in units and units only, which can only be had by the application of points to space. There are elements of time in the mind as the internal sense of time, known as the sense of duration. But we have got to link duration to external cycles to give us usable time in units, as I have explained above.[160] For example, without space we cannot have the year; yet the year is our basic unit of time out of which all other units are derived. This brings a little complication but nothing serious. The reason is that we can only create time, as 'intervals', or 'time units', as I suppose (because the year is only one unit of time and we derive all other units from the sub-divisions of the year with points or mathematics), by means of the application of points to space, thus making time a product of space, and therefore 'space-time'. It makes time necessarily discrete, being the

---

[159] The problem is this: as Sir Arthur Eddington has pointed out, until Einstein, everybody believed that time originated from divine sources---the last hiding place of God after Darwin. If Einstein has now proved that every 'body' has to have its own time, the implication is that time is not universal. But how are we to interpret the having of time by individual bodies? In a word, how do we invent our time? The greatest philosopher alive said it arises as 'relation between points'. For that is how we get the year, which is our basic unit of time. Relation between points means time is derived from space with points—hence it is quite right to call it space-time, so long as it is not implied that the name means space and time constitute one entity in a world of 4-D geometry as Minkowski proposed. This will not cause any confusion because the Minkowski formula is logically invalid anyway.

[160] Duration is our mental facility, obviously born of experience or repeated practice, for distinguishing between the different periods of different time units (hours from minutes, minutes from seconds, and so forth.)

product of points. Therefore time is always 'space-time, or properly, 'space-timed'. That is all the connection between space and time, except that space is required, again, for displaying time in units as we have in the clock.[161] The clock, any clock, does not give 'flowing time'.[162] It merely reproduces units of time programmed into it. The old mechanical clock based on coiled springs gave the best illustration. The springs are manufactured to release units of time: second, second, second. If one failed to rewind the springs, the clock stopped ticking. The springs provided the clock's energy, but were strictly programmed to reproduce time in specific units only.

After the time is derived in this way, it becomes separate from both the space and the points used in creating it. That is why Einstein made them separate entities in special relativity. For, apart from the condemnation of the Minkowski 4-D geometry which assumes that time and space constitute one entity by Russell and Eddington, Professor A. N. Whitehead has also pointed out that time and space still pass through nature separately---not as one entity, in his book The Principle of Relativity. To add to these, I have humbly suggested the Principle of Mathematical Equivalence above, which can also be used to denounce the Minkowski arbitrary and fictitious formula. In science formulas, especially mathematical formulas, are indispensable. But perhaps researchers should try a little bit harder, in the sense that when they are proposing new theories, the old formulas they rely on should be subjected to critical scrutiny just in case their falsifications would destroy their work. In the case of Minkowski, all the Reference Works describe his theory as artificial or arbitrary. Yet it is so crucial

---

[161] As discussed above, the poignant question posed by Bertrand Russell comes up again, namely, in the absence of universal time, what really is measured by the clock? (ABC of Relativity, Ch.4.) This is a very serious matter, because if cosmic time is abandoned, there is no time, or any logical explanation for the time we have. The answer, of course, is that the clock does not measure time. It is deliberately programmed to *reproduce* specific units of time: second, second, second, leading to minutes and so forth, to accord with the cycles of the earth, so that about 31,536,000 (or so many) seconds will coincide exactly with 'one year'. To have more years, we go round the sun again and again and again---hence perpetual time. Units of time in procession give us continuous time.

[162] Consideration of something called 'flowing time' is what stung Sir Arthur Eddington to say some people are making meaningless noises.

for the determination of physical reality. That, probably, is the problem. Things are easy for the mathematicians if space is regarded as four-dimensional. And since there is a formula claiming that it is, well?

SOME MISCONCEPTIONS OF TIME IN RELATIVITY

Some common misconceptions in our concepts of time due to the false nature of the Minkowski equation of space to time in relativity is causing misconceptions in relativity and physics in generally and theorists are urged to speak up against it.

It must not be supposed that the problem of time in relativity has been conclusively settled.[163] Relativity is physics. When a problem is solved in physics the solution is always clear, precise in mathematics and universally applicable; but time in relativity at present is very vague, neither definite nor precise, not least because consideration of time is a philosophical enterprise. The hope is that the original Einstein theory of time can be used to solve the passage and continuity of time. Unfortunately, Herman Minkowski made the question of time in relativity immensely complex and vague, not at all like the original notion proposed by Einstein. Minkowski stands in the

---

[163] This is another of the overlapping essays. Frankly I can't apologise too much for my literary crimes, but I insist it is a small price for solution to the problem of the passage of time!

way. Indeed, as a result, the question of time on the whole is destine to keep the philosophers busy for several centuries as their nostrums become footnotes to Einstein instead of Plato. As regards the physicists and cosmologists, as opposed to the philosophers, they believe that the Minkowski theory makes things easy for them; the problem is that it is just not true of the physical world, and mere mathematical games have no place in physics.

Bertrand Russell has said the concept of space-time is perhaps the most important theory Einstein introduced. To me, there is no doubt (no 'perhaps') about it. It is the most revolutionary theory in human history simply because time is second in importance only to life itself---and yet that life cannot even be lived as a well-organised existence without time. That is how momentous time is in human affairs; and Einstein has shown that it is very different from what it has been traditionally assumed to be. Secondly, he insisted that it should be taken as a separate coordinate in the study of phenomena. In the determination of physical reality, because of Einstein, time is a co-ordinate in its own right just like the height or length of matter and space are, thus making 'Man', the observer, part of the observed, since he has to add the time in the 3+1 formula. Those mathematicians who assume, on the Minkowski theory, that time can be incorporated into space with mere mathematics so that we can dispense with the 3+1 formula and the metaphysical role of man in the determination of physical reality, are contradicting Einstein, which is something approaching a hanging offence in science. On the contrary, it is possible that the passage and continuity of time (or 'space-time') can be conclusively resolved with the original Einstein theory of time as space-time, or local time. (The merging of space with time, which is what we call space-time, as Bertrand Russell has pointed out, was already implied in the special theory of relativity before Minkowski. That is what local time means---time derived from your local space only.) The importance of this simple statement is that it changes everything in the universe of human affairs than agriculture, electricity and the quantum theory taken together.

In my opinion, there is obviously fear in some people that time cannot be something we invent by ourselves. But of course, if 'there is no longer a universal time' we have to find out how we get our time for we do have something we call 'time'.[164] However, nobody is claiming that man invented the whole of time. Rather we have found that we invented how to quantify time by linking the natural sense of time as duration in the mind to external cycles without even knowing the true nature of time. Every method we have ever used for reckoning time has only been able to employ regular or repetitive events or motions that give us only the rates of the passage of time, whatever it is. What is certain is that the sense of duration of anything is connected with the memory mechanism for the retention of images and concepts in the mind; and this sense of duration has built up from repeated experiences over centuries. We have learnt from experience that periods vary.

Let me stress again, and more strongly, that the sense of time is duration in the mind. How long any event goes on in a person's head is time; it comes from or gives the sense of waiting, enduring, lingering or 'time' for short. Professor Eddington also made this absolutely clear in The Mathematical Theory of Relativity; and we have to take that view seriously because the theory of time outlined in this book is based on relativity. Unfortunately, the mental sense of duration is not enough. It cannot give time for general use because it is private; and Professor Eddington stressed this important point too absolutely clearly.[165] For the word 'time' is meaningless until it is objectively quantified. As previously mentioned, there are two aspects of time: the physical and the mental. We need time in units to apply to the external world---i.e. to mechanise it in the clock for general use, so as to be able to tell 'How much time' at a glance. This is achieved with external cycles, the most basic of which is the earth-year out of which all other units of time are derived with mathematics. It is maintained that this

---

[164] It is not often realised that philosophy is of great importance to science; and, as an example, this is the sort of thing philosophers do behind the scenes to make their suppositions indispensable to science in general; for philosophers service every branch of science. The phrase 'survival of the fittest' from biology, which has passed into general usage in science and linguistics, was coined by a philosopher, not Darwin. All the sciences need philosophical interpretations. In the quotation above from Professor Dingle (if you read the full article), he was saying this very strongly in respect of physics; but all the sciences need the same sort of assistance from philosophy, including even mathematics and logic.

[165] Professor Arthur Eddington, ibid, Ch. 1.8.

is in complete conformity with the Einstein notion of time, and therefore incontrovertible. Above all, it is the only means by which we can logically solve the problems of the passage and continuity of time. On the other hand, for the mental aspect of time, we need to know the lengths of the various units of time; that is the only means for telling that an hour is shorter than a year, and so forth.

For now, we are told in all earnestness from the discussions above that relativity is not properly understood. This may be so. But actually relativity is only a theoretical system, a suggestion. It is based on the suggestion that physical reality is not homogeneous but fragmented, and therefore subject to different natural laws of which time is an important component because time is a catalyst in chemistry and everything, the whole of evolution, is chemistry; but time is not universal or cosmic (or something coming from somewhere). It is partly created by us and so it means man contributes something to what we call reality. This is the real revolution at the core of Einstein's ideas, and Bertrand Russell, as usual, was clever enough to notice and comment on it. He calls relativity 'a logically deductive system', meaning that Einstein thought about the universe more logically than anybody has ever tried to do.

In plain language, it is a new philosophy of physical reality so logically structured that it demands attention, respect and serious study. And these Einstein has certainly achieved. With Einstein alone we are not talking about genius but a godlike intellectual phenomenon never seen on this planet before; he reconstructed the world of physical reality single-handed, that is the reason he is indispensable to both scientists and philosophers.

So Bertrand Russell was absolutely right. Einstein's system is a new logic of physical reality, and it works. But theoretical physics is most unlike the physics we apply in laboratories. Ordinary physics is much more like chemistry; it has direct consequences. The Nobel Committee was right to award Lord Rutherford the Prize for Chemistry, even though he regarded himself as a physicist, who had rather cheekily claimed that "all of science is either physics or stamp collecting"!

In theoretical physics there are no obvious consequences, so it is difficult to judge the merits of suggestions. Instead, when we get a new theory in advanced physics (rightly or wrongly), three things will happen. I mean, all three will definitely happen in succession, whatever may be the merits of the new proposal. First, we will get interpretations of the basic postulates proposed in such complex settings (or confused formats) from rival theorists that the debate just has to go on; nothing will be settled in the meantime. But because there are no consequences, nobody will get hurt, no machinery will cease to function; calamities will not occur. The rains will not stop; the sun will not dim.

The most recent example was the eather debacle (or debate). Secondly, we will get accusations and counter accusations of misrepresentations and misunderstandings. The third possibility (because philosophers share with theoretical physic one subject-matter, being the determination of physical reality), will be philosophical interpretations to arrogate the almighty right to shame and discredit some of the factions in the debate, only for philosophers of different schools to turn the tables--- and so the debate will be carried on and on. These philosophical discourses are often pretty profound, giving several intelligent interpretations without being able to settle the argument one way or the other. Strangely, that is how we eventually acquire our knowledge of the external world, sometimes referred to as the practice of 'academic freedom'. That is what happened to Plato. And that is what is happening to Einstein as he has come to replace Plato, in fact, to make his basic suggestion redundant, if not completely false, due to the quantum theory. Newton is already gone!

A careful examination of what has happened to Einstein's theory of time so far betrays elements of all three conditions. First, we are told that 'most definitely' due to Einstein's analysis of 'Order and Simultaneity' there simply is no 'standard or absolute time frame in the universe'. ('Time Frame' or 'Time Reference' means the same thing. It means the logical criterion of validity.) This is generally accepted as true; for it is reinforced by the Lorentz time dilation and local time concepts.

However, it implies that time in the abstract is utterly indefinable, as I have shown with discussions about the earth-year. The year is indefinable; other time units in use on earth are defined in reference to the year.[166] But the year on its own is logically indefinable. Again, all our time units, down even to the cesium units, are based on the earth-year; they are meaningful only as related to the year or fractions of the year, such as the second; but like the year, on their own (that is in the abstract) none of them can be logically defined, because they would not exist without it. In physical reality, all definitions are based on relationships or connections. This is the reason the year, being on its own, cannot be defined, except in terms of space or distance; yet they are not time. Used as an amount of duration it can never be defined on its own. So I feel sorry for all those mathematicians and astronomers estimating the age of the universe in years when we don't even know how long is one year! Or whether it is long enough---which it most certainly cannot be because the orbit of such a small star like the sun does not take much time in terms of the cosmos.

How long, for instance, is a second in logic without reference to something else? The result is that we all have to use the clock, or clocks, based on the earth-year. By this theory of time (as quantified time), the human intellect is built upon the concept of "points and instants". Instants, of course, do not exist independently in nature. Only points do; they had to be discovered by man, but they do exist in nature independently—for example, trees constitute points. Before we learnt to put points on paper, we could see that trees dotted the landscape. Thus points constitute the basic instrument of human thought, especially in mathematics from which all the sciences spring. The instants arise from the act of 'consciously' and 'purposely' moving from point to point, confirming the Russellian notion that time is 'relation between points'. Hence quantified time is human in origin, except that the internal sense of time (as duration of anything in the mind) must be recognised as making psychological contribution to the invention of quantified time in that the external cycles used for quantified time (the years, for instance), have to have psychological anchors (or meanings) which are the senses of duration of anything in the mind. For time, specifically, helps us to know the different periods or durations of the different units of time in the mind to accord with the physical reality of time or physical units of it---i.e. that the second is shorter than an hour

---

[166] See below for more arguments about this issue.

and so forth; so that when they are mentioned, you would know the difference at once.

Secondly, about time dilation for instance, in the absence of a standard time frame, what does it mean to claim that time intervals in a moving frame are shorter---shorter as against what kind of standard or universal time? What time intervals are they compared to since there is no standard time frame? (Note that you cannot say they are shorter as compared to other clocks outside the moving frame; that will bring in the Einstein theory of frames, as I will discuss presently.)

So we all, in the end, have to resort to using the clock or clocks based on the earth-year. Yet if we use the clocks then it is not correct to claim that time intervals in a moving frame are shorter; they are not naturally or normally (in their proper settings) shorter or longer because they are normal to that frame, or to their natural frames. The moving clock may only seem 'different' as viewed from the outside; but if that is the case then there is no puzzle.[167] The time of the moving frame is not 'our' time; and it is not queer to its natural environment or setting. It is a strange phenomenon to those looking in from the outside, in breach of the Einstein theory of frames. In fact, it is irrelevant to anybody but those in the moving vehicle alone. We have to acknowledge that at the time of the Lorentz experiments the Einstein theory of frames was not known. The results of the experiments played a large part in helping Einstein to conceive the theory that different frames in nature obey different physical laws, which, as I have said before, is one of the most profound philosophic insights in history.

Now that we know of inertial frames, I find that the whole idea of studying other frames from the outside is fraught with difficulties; it can never be an exact science since the standard postulates that make our system work (and make it what it is)

---

[167] Otherwise it is difficult to see how the behaviour of one clock can affect all time, human physiology and even the material contents of atoms, e.g. muons. If time is defined as the passage of existence in consciousness, how can the behaviour of one clock affect it for all of us? There are still a lot of religious beliefs about time. Time dilation is one of them, so sweet to the religious in science because they can claim that "it is a unique mystery about time predicted by Einstein". In fact, it is not a mystery, let alone predicted by Einstein: he rather solved the little problem with his theory of frames—i.e. the dilated clock belongs to another frame to which it is running normally.

might be inapplicable outside our frame, or planet.[168] Speculations into other frames from our frame have been responsible for all the bizarre suppositions about time and space-time from mathematicians and cosmologists in general relativity. I do not think that kind of enterprise is justifiable, especially when it leads to theories that space-time may be infinite in its timelike directions.[169] Space-time cannot be infinite because it is necessarily discrete---the year, for instance, is not infinite. It is only one; all other units of time derived from the year are also discrete and individual. The proper way to think of time as space-time is that its units are in perpetual procession (one year or the second following other years, seconds and hours successively) to make time seem to be continuously passing by; as such time can never be infinite. If the source of the units is cut, the procession would cease; that is to say, when the earth disappears our time will end. The concept of infinite time is religious in origin. Minkowski used mathematics to try to cement it in science but unfortunately his mathematics must be based on a logical premise, and that he could not provide. What he provided was imaginary and therefore unacceptable.

Now I come back to muons which I mentioned a few pages back. Nothing illustrates the confusion about time in physics as a result of relativity and how it is misunderstood by scientists than the story of muons. By normal logic they should not last long enough to reach the earth; but they do. With the use of formulaic mathematics and concepts, physicists explain this by saying special relativity provides the answer as follows: the

---

[168] I think one implication of this is that the laws of physics, or some of them, would differ from ours at least in some parts of the cosmos, if not all over. Einstein was really a very strange genius in physical thought. He introduced the notion of postulates for natural laws in frames. This idea may go very far indeed in the cosmos at large---i.e. physics may not be a uniform theory applicable to all the cosmos; it may depend on the nature of the postulates inherently present.

[169] This is the view put over by the august Encyclopaedia Britannica. Most cosmologists (it states in the Macro) now assume that space-time is infinite in its time-like directions. Hence there are massive tomes dealing with the passage and direction of time theoretically; yet all the time the units of time (the year and its fractions) are passing by through the procession of the units---second, second, second, hours, minutes, and years, all are passing in procession to make time pass by without theories. .

speed of muons is so great that their internal clocks slow down. Using the theories of time dilation and the so-called twin paradox based on it, it is assumed that as the muons speed and their internal clocks slowed down they aged less and thus are able to last long enough to reach the earth.

To a logician or philosopher who understands relativity, this is so laughable as to choke him. It is really the best example of the confusion in physics about time in relativity. (1) Time dilation has nothing to do with the muons and how they behave, since time does not dilate at all. Lorentz found that a moving clock would be seen by outsiders as running slowly; but internally those carrying the moving clock would notice absolutely no difference in its performance. Einstein explained this with his theory of frames---the moving clock is in a different frame. There is no logical mechanism for this kind of episode to be able to control all time per se. All other clocks would not run slower or faster; and since there is no such thing as an absolute time frame, or a standard time, by which all other clocks can be compared, the moving clock's performance has no relevance at all in physics, because its carriers would notice no anomaly; and those outside who notice any anomaly should mind their own business since it is not their time. (2) The idea that muons have internal clocks is based on the Minkowski theory of space-time, where space and time are assumed to constitute one entity; therefore the reasoning goes that, since the muons occupy space, and all space is space-time, they have their own internal clocks to keep or measure time for them. Again, any logician will describe this as nonsense; for after all, the Minkowski space is known to be fictitious and arbitrary with absolutely no logical validity. Secondly, the very idea is easily disproved thus: we know there are (roughly accurately) specific times on our normal clock for certain events on this planet---let us use Sunrise and Sunset for illustration. If Sunrise is usually 6 am and Sunset is roughly 6 pm as they are in some countries in the Tropics, it is inconceivable that a moving clock can force or influence these times to become 5 am and 5 pm, on the planet just because any clock is running an hour late.

The reader will have noticed that the name of Lord Bertrand Russell comes up regularly in all discussions of relativity's interpretation. It is inevitable. Russell was highly respected by

Einstein, and for very good reasons. He was the world's greatest philosopher at the time. He was also a great mathematician and logician of genius. A most attractive writer, who won the Nobel Prize for Literature, he wrote about every subject in philosophy, including novels to illustrate moral points. When relativity was announced, he abandoned many of his most cherished ideas as wrong without shame or even mild embarrassment. He was candid and honest in the most adorable way, completely dedicated to the truth no matter how it reflected on his own beliefs. Russell probably had no certain beliefs other than the pursuit of the truth wherever it took him: via science, logic, mathematics or plain common sense, and linguistics. If he were certain that teaching mathematics to people from the cradle could save the world, he would have advocated that as his philosophy.

Concerning relativity specifically, in the later editions of his little book "Problems of Philosophy" he denounced his original philosophy as expressed in the book because of Einstein's theories, joking that whoever wrote the original ideas must have been a monkey, but nobody should suppose that the monkey looked, even remotely, like himself! No great philosopher has ever made such a confession. Often associated with rulers, they all wrote imperious edicts as if they had discovered the final truth in logic and metaphysics.[170] Indeed, Russell later called his Fellowship dissertation

---

[170] No surprise, then, that Russell later put them in their deserved places (mostly of dishonour) in his monumental *History of Western Philosophy*. One complaint is that he never even once mentioned the name of Wittgenstein in this great book. The reason came from his contemporary, Sir Karl Popper---it was because (and I am repeating this for emphasis), "In the long history of philosophy there are many more philosophical arguments of which I feel ashamed than philosophical arguments of which I am proud...Russell saw these things in that light, and so did I..." (From, *Modern British Philosophy*, By Bryan Magee, Secker & Warburg, London, 1971.) In 1959 Russell published his book, *My Philosophical Development*, in which he said he eventually had to reject Wittgenstein because he was preaching 'logical mysticism', which was anathema to his basic nature. Of course he was right. Correctly defined, logical mysticism includes religion, mysticism and unscientific gibberish, all dressed up to look like valid logical reasoning with a variety of linguistic trickery. Many aspects of philosophy in Oxford and Cambridge (and elsewhere) remain stuck in this kind of sticky intellectual mud ever since.

"somewhat foolish" for the same reason, namely, the geometry used by Einstein had made his discussions of the foundations of geometry completely wrong, and he was happy to admit it. He wrote one of the best interpretations of relativity, still in use, under the title "ABC of Relativity". His book "The Analysis of Matter" can be divided into two. One section is about relativity; the other is mainly about his joint theory with A. N Whitehead to the effect that the world of sense is a construction, not an inference.[171] Yet even this can be traced to relativity, since Einstein made man the observer part of the observed, meaning that man contributes something to the nature of physical reality---i.e. to help with the construction of that reality---and the book was published long after both special and general relativity. It is a moot point. For the Einstein theory used the 3+1 system. The three facets of phenomena are natural; the time is, in Einstein's system, one's own local time. It means one would have to invent his time as a union between the sense of duration and external cycles before having an "objective time for general use in one inertial frame" to add to the three natural dimensions of phenomena, to complete the construction of physical reality---or the physical reality relevant to one's frame of reference.

---

[171] This has now been proved in physics with the theory of QED.

12

REPLY TO SOME CRITICS ON THE WEB

Those writers on the internet who think Minkowski is God because his theory makes time travel plausible are urged here to stay away from philosophy. I see the internet as a marvellous, even great, invention, hijacked by evil men intent on destroying everybody's peace of mind and even the world. I am here answering the writers on the internet objecting to my theory; but I would suggest that the web should be strictly controlled and even placed under the surveillance of the security service.[172]

In the interest of learning, I am sending you the Appendices of my recent Monograph in which you will find all the answers you seek about my work. However, please understand that I am presenting a rounded philosophical theory about time, how it

---

[172] This piece and the next chapter (just going over the arguments in the book) were posted on the web in 2006. It is included here (after some redactions) in the hope that some of the points raised may be helpful to readers. But inevitably they overlap. To be charitable you could say they are complementary, although I have added it to my literary crimes. I am too old to fear punishment!

passes, and how it seems continuous in an attempt to solve the problem of perpetual time without the involvement of God---all of it based on Einstein's notion of time so as to link philosophy to physics by means of time alone. For a very long time I have felt that it has become possible to do so, either by time or by means of the quantum; particularly the quantum because it is the same thing as the light by which we see things.

To be completely rational, epistemology can never ignore the quantum theory since the quanta by which we observe the world are also matter or small pieces of matter.[173] It cannot be ignored in any theory of physical reality, however it is conceived. To link it to philosophy is to abolish the philosophy that regards physics as 'just another way of looking at the world', and see it instead as the only way, rationally; for the quantum which is the basis of physical reality is what we see as light and which leads us to have vision of other material bodies. Even the Platonic theory of Ideas becomes redundant because outside the quanta images cannot exist; and the quanta are seen plainly as light---so we see how images are constructed, and by what means. By the quantum the demise of Idealism is finally concluded.

Also, because the quantum is time-dependent (as 'energy-second'), I have made one or two comments about it. I don't think it can be the same throughout the universe because the time by which it is known on earth is peculiar to the earth---that is, provided all the universe is not subject to the 4-D geometry of Hermann Minkowski, and therefore a second here is not the same as a second everywhere else in the cosmos, according to the original Einstein theory of time. Thus refuting Minkowski is crucial. As energy-second, the quantum's energy is natural---the time is not. It is our peculiar second; and I have discussed how we make our seconds on this planet at length in the book, suggesting that it could have serious implications for the Theory of Everything, too.

---

[173] Why can we only see the external world by means of quanta when the external world also consists of material objects created by the congealment of the same quanta according to QED? That is the conundrum introduced into human thoughts by the quanta, which are part of Einstein's scientific ideas. At once it makes the Platonic (unscientific) Theory of Ideas redundant.

On the nature and passage of time and 4-D geometry

The nature of time may have a bearing on the Theory of Everything due to the following observation about time. First of all, since Einstein was not a 'professional' philosopher he did not attempt to give the logical grounds why every separate body in the universe has to have its own time.[174] Unlike mathematics logic is mercilessly dry, acute and uncompromisingly factual; everything must be clearly defined; all conditions and methods clearly spelt out. Logicians cannot invent their own rules as they go along like mathematicians.

The Einstein theory of frames is used to justify the claim that time is limited to a frame, but the technical grounds why this is so have never been made clear. That is to say, the conditions in nature that make time limited to a frame have never been clarified. Einstein could not be blamed because he was not writing philosophy. The philosophers have failed to interpret his ideas properly----everybody complains that relativity is impossible to understand.

Let me state the 'necessary logical grounds' why time is limited to a frame in a clear language (without mathematics) for the benefit of the reader or readers: time must be quantified to be useful in science and logic---let us say 'to make it usable culturally'. It is meaningless to just mention time as such. Culturally we can only use quantified time, otherwise how could we mechanise it in a clock? Now, to quantify time we have got to employ external, repetitive cycles (or regular motions)[175] in association with the internal sense of time, which is the sense of the duration of anything whatsoever, to get our usable periodicities---or time in units; so that the duration is converted to time units, or usable time; until then, time (as duration in the mind) is not usable, and can be sensed only in the passage of

---

[174] Note that Einstein qualified for the noble title of 'philosopher' for a number of reasons. For instance, the difficulties over the eather or the propagation of light arose simply because the nature of physical reality had been misconceived by both physicists and philosophers---this went all the way to the quantum theory. Only a great philosopher could have solved such problems and make the solution part of mathematical physics, and not as an unproven supposition, or suggestion. I will give him the title of 'The greatest Thinker', not the greatest scientists. That honour belongs to Charles Darwin, no matter what the religious people say against him. Life is superior to everything else.

existence, motion or silent ageing. These quantified units of time then become unique and applicable only to the body concerned, the body whose cycles were used to create them; it is from that body's cycles the units were established and so they could not be appropriate to any other body 'without ambiguity', as Bertrand Russell observed. Using complex mathematics to overcome the discrepancies is an act of human interference of nature. At some stage we will come face to face with insurmountable quandaries due to the misuse of time via the Minkowski fiction.

This is not nit picking in logic because on the Minkowski interpretation of time, all the works of cosmologists in the supposed metric of general relativity is vitiated. I believe that attempts to conceive a Theory of Everything has also suffered from the use of earth time everywhere, when it is obvious that it cannot be applicable everywhere: we want to link the quantum to gravity. Yet the quantum cannot be 'a universal unit of energy' because it is time-dependent as 'our' energy-second. This must be taken into account, but it is not. Scientists just use the word 'time' and forget about its quantification and unique periodicities tied to the earth. It is foolhardy to believe that a universe as huge and complex as it is can be subject accurately to the periodicities derived from a tiny object like the earth.

To get a fair idea of my supposition you will have to read my books about my theory, of which there are more than one. Failing that, these Appendices to my latest work will give you an idea of my philosophy. Please note that those aspects of the Minkowski mathematics you cited have no logical merit. I am questioning his basic premise. I insist that, for time alone, the i in his ict equation is not tenable; therefore his 's=ct...' deductions based on the ict equation are wrong.[176] There is no such thing as 'imaginary time'.

---

[176] At all times it should be realised that, despite the condemnation of some scientists, philosophy is important; believe me, it is very important. Einstein said he was influenced 'very greatly...' by David Hume and E. Mach. To get at what can be considered as the ultimate truth (or the truth for short, if you like), philosophers, unlike mathematicians, have to go to the roots (the logical foundations) of equations; merely repeating the mathematical symbols as written is regarded as shallow, at this level, even an insult. Let me quote part of what Russell wrote about the Minkowski formula---and you

Mathematicians often forget that mathematical symbols must have causative meanings; but philosophers never forget that. It happens to be one of the obvious differences between philosophy and science. Statistical mechanics in science (as opposed to direct one-to-one causality) overcomes the quirks and deficiencies in the behaviour of phenomena due to the absence of direct one-to-one causality in the nuclear and sub-atomic matter; but that does not mean the old philosophical causality can be dispense with altogether; for causality still occurs, only statistically. So, logically, statistical mechanics is also caused. It may not be as direct as throwing a rock to shatter a glass window; it is more like your rock going through intermediaries before reaching the glass window, so that crooked lawyers can disclaim liability; but in logic you're liable for indirectly causing the damage. This is a brief account of the type of causality now envisaged under statistical mechanics. The many mysterious behaviours of sub-atomic and nuclear matter are not without cause; for they occur because those particles exist; if they did not exist, the events associated with them through indirect causality (or statistical mechanics) would not occur, as is shown in QED. There is so much in physics crying for research in dept, which scientists have neglected by relying on the fictitious Minkowski formula.

Take the quantum for example. (As energy-second, the quantum is time dependent; it materialises periodically in accordance with our time, the units of our time as explained in the section about quantified time. If this time is peculiar to the earth, as I think, then the quantum mathematics cannot be universally applicable; and so the fear that the cosmos contradicts the law of direct causality might be misplaced.[177] I could not put it stronger than that. It is sufficient to indicate by this idea that more research is needed, as I suspect that the energy-second which is applicable on this planet might not be

---

cannot say Russell did not understand the mathematics of Minkowski: "…the philosopher cannot but feel dissatisfaction with the apparently arbitrary assumption about intervals…" And, again, "…there is great difficulty in suggesting any non-technical meaning for interval; yet such a meaning ought to exist, if interval is as fundamental as it appears to be in the theory of relativity…" (Bertrand Russell, *The Analysis of Matter*, Ch. Xxxviii.)

[177] Einstein may turn out to be right after all about this matter.

universally applicable, and so we cannot rely on the nature of "our quantum" alone to argue that direct causality is cosmically abolished---it may be so from our point of view only, for after all, how important are we? There are stars so immense that millions of our sun will find room in them. Also, we must think of the size of our planet as compared to the sun---and the size of a human being in all that. I suspect that the quantum is not the end-piece of matter or energy in the universe at large, as opposed to what happens on this minuscule dot of a home for man.)

Back to Minkowski, he cannot hide behind the obvious lack of direct one-to-one causality to try to alter physical reality with his counterintuitive mathematics. Knowing time as it is, where is the imaginary time coming from, and what is it supposed to be like? In other words, what is the meaning of 'imaginary time'? Time, once you think of it, ceases to be any other thing than time in the clock, or quantified time as I have defined it. In quantified time we accept that time does not physically exist out there as a separate and identifiable entity like rocks and rivers; but we do have something for use as time in society; and analysis of that something shows that we merely count cycles as the rates of the passage of time---even without knowing what it really is.

The concept of imaginary time (if the reader is not aware) was invented by Hermann Minkowski; that is the premise of his ict mathematics purporting to equate space to time with counterintuitive mathematics; the i was meant to invoke imaginary time. The idea is arbitrary and therefore logically untenable.

13

WARPED SPACE AND CURVED SPACE-TIME

Warped space excites the internet popular imagination because of time travel. This imagination is rapidly becoming erratic and paranoid and totally devoid of any rational inclination for logical debate. They are warned that, although warped space is real but since space is not naturally linked to time as Minkowski proposed (and therefore space and time are separate entities), there is no such thing as "Curved Space-Time" by which time travel is assumed to be 'a scientific possibility' by some writers to feed the geeky imagination on the internet. There is curved space, of course. It is just about the only thing proved in general relativity. And it is true that it can curtail cosmic distances; the example usually given is like folding a piece of paper to make both ends meet at once without using the intervening space; but if time is not in space naturally then it cannot take time with it when it folds or curves.

This matter which is also in response to queries from the internet deserves a sub-section (a chapter) to itself because all scientists seem to have fallen hopelessly in love with it, quite wrongly, I think. Of course, we all know that the Einstein notion of gravity as caused by curved space has been proved. Then Minkowski came along to claim (merely claim) that space and time constitute one entity, and therefore when space curls, time is also curved with it. That idea is arbitrary and false because his ict equation upon which it is based is logically flawed. It is simply not true that time is intertwined with space and can curve together with it so much so that you could (using the appropriate mathematics) meet your grandparents even before they were married. That is pure mythology, and comes from the Pythagorean 'Trans-migration of souls' long since discredited. If it were true none of us who are not millionaires would still be living on this planet; for wherever our grandparents might be we would join them; anywhere is bound to be better than this world!

More seriously, you will find that I know the Minkowski mathematics pretty well to even incorporate it in my corny jokes. At this level every writer is a mathematician of sorts. I even agree that he makes relativity easy to understand from the point of view of mathematics—i.e. by dispensing with the 3+1 formula and still have time inherently in space as a separate co-ordinate. He could do so because time cannot be suspended and can be invoked with any symbol; in this sense, and since his basic logic is flawed, you could say that he was an intellectual fraud. On the other hand I am not also trying to invoke time with any symbol; I am not an intellectual fraud. I am rather trying to show that any method for demonstrating time can only show how it is passing and never what it is; therefore what we call 'time' is how it is passing and need no other theory to explain how it passes by. So you could say my definition of time is pragmatic---namely how we experience it not what it is; but then how we experience anything is what it is in so far as we are concerned! This is a better definition than claiming that it is part of space by means of flawed logic so that the theory is called arbitrary. The question is why do scientists use the Minkowski formula knowing that it is logically flawed?

Please (always) remember that Professor Eddington said, although useful, we must never forget that the Minkowski theory is fictitious. To me even that is unacceptable. It is not strong enough for me as a condemnation of the Minkowski proposal; useful or not, what is fictitious has no place in physics at all. It is distorting relativity so much that it is still not properly understood---see below.

It really is amazing that experienced and serious scholars and scientists are discussing time as if it is something permeating the universe; and the Americans in particular state their theories as if they are part of established science. Yet, by an elementary logical deduction, it is easy to establish that "There is no longer a universal time under relativity", if time is limited to a frame, so much that every inertial fame is supposed to have its own time system, and, therefore, in the words of Einstein, "There are as many times as there are inertial bodies." Our obligation is to trace how our time began, how it passes by and also how it seems continuous. At least the atomic time indicates that we can only use regular or repetitive motions to show how time is passing by---which is, in my opinion, all that we can ever know of time. I have mentioned the atomic time, but the earth-time is similar. Orbits, pulses and oscillations of anything used to track time can just show how the time is passing only, namely how many oscillations are needed to do so-and-so, say, for boiling an egg. So when the necessary numbers of oscillations pass, you know that your egg is done. But that is passing time. It is what we record as the passage of time. The pulses have to occur before they can be counted; but they do not stand still---they are passing. You know that your egg is done after so many pulses have passed. Tapping the finger gives the same results. I think I have made the explanation of time and its passage simple.

My argument is that it is all we can ever know of time and never what it really is. Nobody knows (or can ever trace) the nature of the phenomenon by which an egg is cooked after the lapse of mere physical pulses of an atom. I have mentioned chemical processes, but like everybody else, I am only guessing. The nature of time can never be known by man. Process and reality is the thought most closely used to interpret a period of waiting---which we know as time. The process may be chemical in nature; but we experience it as a period of waiting, or time. This phenomenon is what scientists are trying to explain with the notion of imaginary time. They say that time is linked to space.[178]

---

[178] It is true that Einstein himself proposed this idea of time after the Lorentz discovery of local time. That is why he said he was proceeding on the theory

This time is therefore called 'space-time' and is defined as a mathematical space existing by means of three space coordinates and one of time---how do you get that time without using repetitive cycles that make use of space? The answer, we are told, is that we do not get any time. The time coordinate is assumed to be imaginary, which makes me want to throw up! How does an imaginary entity come to be physically linked to, and become an inherent part of, space? This is absurd. All I can say is this, if the mathematicians think they are fooling us, they cannot fool nature and therefore very soon physics is going to have to put its house in order.

Almost all the work done in theoretical physics and cosmology since Einstein is adversely affected by this notion of four-dimensional continuum---simply because it does not and cannot exist, except in mathematics and in the imaginations of those who propose it. They plead that they cannot understand relativity without the imaginary time coordinate idea; but it is better not to understand a theory than trying to distort it with pure fiction. Time is not fiction. Time is physically tracked through the use of repetitive or regular cycles; and that is how it is passing by. I concede that physically we do not know what it is. But it is physically there because we can feel its effects in ageing, ebb and flow, periods of waiting and so forth. That is the limit of human knowledge of time, and even that is vague enough. Imaginary time is arbitrary, as Russell and Eddington have said, and ought not to remain part of physics.

The important point is that the orbits or oscillations are "relations between points". The year, for instance, is motion from one point to another point. In between the two points we have the length of the year. The pulses of atomic time are logically the same only shorter. These orbits and pulses can

---

that the Lorentz local time is 'time, pure and simple'---the first time in history that anybody had proved that time was (or is) not absolute or generally permeating the whole cosmos. But Einstein used the 3+1 system to include time in space as the fourth dimension. It is easier to assume time as already inherent in space in an imaginary four-dimensional space and dispense with the 3+1 system. But it is impossible to do so unless you rely on an imaginary time, in the sense of assuming that by some magic the time is there already when you have not yet created it, because Einstein and Lorentz have proved that it is not universally permeating the cosmos but that every inertial body creates its own time.

only give us discrete time, being the product of points. It is shocking to discover that this simple deduction is not understood, and serious scientists and philosophers are still talking of time travel backwards and forwards, and also of "the space-time continuum" as if time is continuous or permeating the universe. How can you travel in time when the time is produced instant to instant only? At the risk of repeating myself much too often (my excuse is that the subject is pretty difficult. I feel for the reader!), we have hours, minutes and so forth. Those scholars writing about the direction of time forget that discrete time has no direction---the units replicate to pass by. Instead of direction we get procession of the units.

Since time is added to phenomena in the 3+1 formula because time is a catalyst in chemistry and all creation or evolution is chemically caused, as a supreme act of rare intellectual honesty (not common in mathematics), I have to admit that, in a way, Hermann Minkowski was very nearly right. I can see why mathematicians adore his formula, especially after Einstein used it and even praised him---for scientifically Einstein is virtually God. For if time is always there (even silently since we age 'over time' silently), and it is also a catalyst in chemistry while everything, too, is chemically caused (physically and organically), then it is right to think of time as an imaginary entity that will always be there. And since it will always be there the equation will hold good. For this reason scientists believe in the Minkowski theory. The problem come from philosophers, and philosophy is always important, and it is this: you have to demonstrate this mathematically and the only way is by means of imaginary time coordinates, but here Einstein stands in the way: in the absence of universal time where is the time coming from silently or otherwise? So we have to return to the "If time is always there" proviso. The 'if' becomes critical. How is it there? If you say it is always there how do you know that in the absence of a universal time permeating the whole cosmos and the same

everywhere? You can only do so by equating time to being---but time is different because it requires points. Time means from when to when. So in the absence of a universal time you have got to bring in man's creation of time; and I have found that the only way we create time is by the use of repetitive cycles, hence quantified time---you have to bring in conscious human action, and if so, then time is not always there because it cannot be equated to being. Therefore since time is not always there, Minkowski could not get his time coordinate and had to rely on imaginary time which vitiates his formula.

With regard to the passage of time, let me stress that every unit of time replicates to advance. Also how can time be present in every space as a four-dimensional continuum, or 4-D geometry, when the time is discrete---produced unit by unit, like the year or the atomic pulses? How can space-time that is supposed to be dependent on imaginary quantities be part of physics? What is imaginary time? Is it something you imagine to be existing while knowing that it does not physically exist? Can that be part of the calculations that can physically land us on the moon? It may be useful in certain mathematical proposals, but are such proposals a true reflection of physical reality? Why did Professor Eddington describe it as arbitrary and fictitious? Why did Russell also condemned it? Why does everybody refer to it as artificial? How can you have imaginary space, as space-time, specified by three space coordinates and one of time---where is the time coming from since under relativity time is a product of space? How can space be used to define time as 'space-time', which is supposed to be a real physical entity specified by three space coordinates and one of time? Is this not senseless tautology? These questions are, of course, ignored by the great and mighty in science, including the administrative minions at the Royal Society who refuse even to acknowledge letters addressed to the society.[179] Every institution is always as good as the people who are employed to run it.

---

[179] These questions were fired at a nerd on the web who deserved it.

On the nature and passage of time and 4-D geometry

In the mean time, to see the importance of this matter let me digress with a brief mention of something that I know is worrying mathematicians. In discussing time rationally in that peculiar sense of physical reality championed by Ernst Mach (rather than as 'philosophy' or 'mathematics')[180] I declare that I have nothing against mathematics and mathematicians. In defence of Ernst Mach, let me say this: it is conceded that there are several aspects of physical reality (or science in general) better described (or written) with mathematics. Some things cannot be understood at all without mathematics. For instance, without mathematics we could not have time as we know it, because we could not state 'how much time' in numerical units; and without that the clock would not exist; also civilization would be primitive. This could be a subject in Sci-fi novels---a people without clocks, and therefore condemned to live too close to nature. However, mathematics should not be allowed to dominate the entire field of the human intellect for the simple reason that you could not demonstrate or write what we know about physical reality by mathematics alone; even if you could do that nobody would understand you. The linguistic philosophers believe that language is the most important human medium for serious thought and communication; mathematicians also claim that without mathematics the secrets of nature would remain hidden. You could not tell that an aircraft could fly to Australia or anywhere else without mathematics. Both are right; but not all hidden secrets of nature are beneficial and in any case you have to be able to explain it to people.

The suggestion I am making is already graphically illustrated by the life story of the British Nobel Prize winner, P.A.M. Dirac, the man who averred that Albert Einstein was the greatest scientist

---

[180] Mathematics is necessary for creating time in units (the year, for instance, as resulting from a point to a point; and there our time ends, unless we orbit the sun again); but time can never be geometricized because it is not entirely physical. The physical aspects are used merely to quantify it; but there is the inner sense of time, as the sense of duration---and how do you geometrize that? Feeling the sense of duration is as important as the hand of the clock. If the second hand of your clock moved ten paces at a time, you would sense at once that it was not giving you credible duration of time because we have the sense of the correct duration in the mind. We take the sense of duration and sub-divide it with external cycles to get time in usable units. That is the end of the relevance of mathematics in the study of time.

of all time because, "Only scientists like Niels Bohr and Max Planck were qualified to wipe his boots. His theories came out of the blue. They did not follow from what had gone before" [and, I would add, yet they work.][181] Dirac was said to be the greatest British physicist since Isaac Newton. However, he was regarded by his peers as a poor communicator, and sometimes even incomprehensible. "...as a thinker he was unintelligible except to mathematicians. Even his fellow physicists complained that he worked in a deliberately mystifying private language..." His reply was that, "The quantum world could not be expressed in words or imagined." (Taken from John Carey's review of THE STRANGEST MAN: The hidden Life of Paul Dirac, by Graham Farmelo. Published by Faber, 2008.) Here we have both sides of the argument sufficiently elucidated. The scientific genius wanted to communicate mainly by mathematics. His equally brilliant peers objected that sometimes even they could not make out his meaning. And his response was interesting. He claimed that he was not to blame because the quantum theory which occupied him was necessarily abstruse---and we know he was right. Einstein being 'Einstein', the special theory of relativity was also difficult; and general relativity almost impossible to comprehend by normal scientific thought. So both the genius and his critics are obviously quite right. Quantum theory is abstruse, no doubt about it. According to Niels Bohr "Whoever is not shocked by the quantum theory has not understood it." On the other hand, the genius has a duty to make himself understood, otherwise why bother to communicate at all? In parts of his ABC of Relativity, Bertrand Russell warns the reader not to try to visualise what he was describing in general relativity. That is one way of solving the impasse.

This means that what you state with equations (as Bertrand Russell always did) must be rendered in words too, however imperfectly. If you cannot do that your theory will never be able to stand logical scrutiny due to the absence of clarity in definitions. Here is the example I am most fond of: before Minkowski space and time were separate entities. In special relativity "they play distinctive roles", yet it works. How did they come to be one entity after Minkowski? To dwell on his so-

---

[181] I am quoting from fallible memory but it appeared in his Obituary in either the Guardian or The times some centuries ago!

called 'counterintuitive mathematics' raises two questions: (a) His mathematics must be faulty, for obviously special relativity works pretty well. (b) Mathematics alone cannot demonstrate the nature of physical reality; thus the Minkowski formula does not accord with the physical reality revealed by special relativity. Hence Mach was right.

I set the Lorentz Transformation of neighbouring co-ordinates (formula) aside so as to avoid the impression that the Minkowski ict equation is based on the Lorentz Transformation and therefore acceptable. The common mistake of mathematicians has been to use the Lorentz Transformation of co-ordinates as the basis for deriving the Minkowski formula for equating space to time. This led even the great Einstein to say, "...for to every event there are as many 'neighbouring' events (realised or at least thinkable)"![182] For Goodness sake, equating space to time will never be logically feasible because the time cannot be had without using space, otherwise how do you get it in units? Without space how do you get the year, for instance?[183] The greatest logical mind in science was completely misled by the "Neighbourhood Co-ordinate" formula. Yet there is absolutely no logical method (or route) by which neighbourhood co-ordinates can lead to a natural union between space and time when the time is derived from space in the first place---unless one relies on the Minkowski mythology by supposing that, as Einstein put it, 'it is thinkable'! I do not suppose for one moment that scientists and mathematicians are unaware of the anomaly; surely everybody can see that 'thinkability' is not part of objective reality. I rather think they do not know what to do.

---

[182] From RELATIVITY, op. cit. Part I, Sect. 17. I don't know what Einstein was thinking of when he wrote this. It is an absolute bloomer! However, he based it on the transformation of co-ordinates, and so he felt it was right, mathematically.

[183] How do you get any unit of time without using space? And once you get your unit of time, using space, how do you put them back together to constitute one entity---with the use of incomprehensible, abstract mathematics? This is a case where the Mach doctrine against the excessive use of abstract mathematics in the study of physical reality is particularly relevant; we have always to remember how 'greatly' that doctrine helped Einstein.

These problems arise for the so-called interpreters of general relativity because they have simply got the definition of time wrong---if they have any at all. Most of the time all scientists and the general public still believe that time just happen to exist. It was this lack of clear definition of time that led me on to the insight that it actually cannot be defined because it is not a physical entity we are intellectually equipped to be able to identify. We can trace how it is passing by through the use of repetitive cycles, pulses or oscillations of atom, but nobody can ever define its real nature. Even trying to define how long is one year is impossible in temporal terms, except in its distance. Fortunately, knowing that we can only trace how time is passing by helps to solve many of the quandaries of time. In other words, there is something we use as time; but we get it only through repetitive cycles and can never know its real nature since it is not physically existing out there. But these repetitive cycles give us discrete time and that is the critical point added to the notion of 'quantified time', or how we get time in units for cultural use even without knowing what it is--- one of the greatest mysteries on earth, or perhaps the greatest of all mysteries. I don't think the mystery of time can ever be conclusively resolved to everybody's satisfaction that it is purely secular. If it is so, then why is it closely associated with life since we cannot understand the latter or whence or why it came to be?

On the other hand, mathematicians thought the Minkowski formula was a blessing as it makes things easy for them by incorporating time into space and therefore easy to define as consisting of special intervals; in fact, it has turned out to be a curse, and a very serious one, too.[184] The 4-D geometry is the ideal solution, or rather I should say, it would be the ideal solution if it were true of the natural world. Since it is not true

---

[184] Bertrand Russell put it best: "...the philosopher cannot but feel dissatisfaction with the apparent arbitrary assumptions about interval..." and also, "...there is great difficulty in suggesting any non-technical meaning for interval; yet such a meaning ought to exist, if interval is as fundamental as it appears to be in the theory of relativity."---The Analysis of Matter, Ch. XXXVIII. By this time the Minkowski formula had been incorporated into general relativity otherwise it was not originally part of it, and, of course, we know now that Einstein never really accepted the concept of space-time continuum as logically valid.

and therefore is untenable in physical reality, it belongs to the realm of fantasy---'Dream Physics', I call it. So they do not know what to do, because they have already incorporated it into physical theory; they can do that, as I have said before, because in theoretical physics no immediate consequences flow from theories[185]---therefore nobody gets hurt when theories go wrong.

Now mathematicians must swallow their pride and agree, as Kurt Gödel argued, that whatever we do certain aspects of mathematics can never be completely objective.[186] This is a clever notion; and it accords with the Platonic simile of the cave. Thus, in relativity, we must revert to the 3+1 formula used by Einstein in special relativity---if it worked there, it will work anywhere else due to the 'Two postulates'. 'Anywhere else' means in any inertial frame subject to the 'Two Postulates'. For after all we need physics (or theories of physical activity) to be effective only in an inertial frame, which is the field of special relativity applications, and where we know that the 3+1 formula works absolutely perfectly, not in the metric of general relativity where we know that human life does not and cannot exist---so what is the need for time in general relativity?

The 4-D geometry is merely 'assumed' to work in general relativity without proof. I actually believe that most of the post-relativity work in general relativity and cosmology has been falsified by the reliance on the Minkowski formula---but mathematicians only have themselves to blame, because Bertrand Russell and Professor Eddington said plainly that the Minkowski theory is fictitious and arbitrary, which meant that, logically, it was only a matter of time before it would be rejected by thinkers.

---

[185] No such thing as touching a button to produce results.

[186] Incidentally, I wish to point out that the Gödel formula, known as the Gödel Universe or Gödel Universes, is vitiated by the Minkowski theory upon which it is based. Every supposition influenced by the Minkowski fiction is bound to be logically flawed. It seems to me a sheer waste of intellectual effort since Eddington, Russell and Professor Whitehead told us the Minkowski formula is logically untenable. Perhaps scholars have been encouraged to rely on Minkowski because it makes things easy for them, and because even Einstein used it. But I am convinced that Einstein was coerced. He felt the need for support. Secondly, he knew that it could not affect relativity.

Yet they did not listen. Objections were regarded as evidence of one's ignorance of counterintuitive mathematics. All the time the proper definition of time was not even attempted. For example, where will the next second come from if the earth stops orbiting the sun? From the obvious answer that it cannot happen (or at least not just yet!), because the earth is gravitationally programmed to always go round the sun, we get the evidence that (a) the time units we use to tell 'how much time' without which our civilization could not survive, are derived from the repetitive motions of the earth round the sun; and, as such, they seem clearly human in origin. That is one obvious proof that we use external cycles in union with the sense of duration to quantify time for cultural use. (b) It also goes to show that the continuity of time is obtained from the succession of time units---the repetition of the yearly cycle with its associated fractions, for instance.

The year is what we sub-divide to get all other units of time on earth. Thus no matter what mathematics is used, it is not possible to equate time (derived from space) to space again! That is a contradiction in terms. The whole of post-relativity physics, the real nature of physical reality, and even relativity itself are all distorted because of this mistake by mathematicians.

I rather accept the contrary position taken by the Mathematical Society of Japan, from whose Encyclopedic Dictionary of Mathematics[187] I quote the following consensus: "Historically, the transformation formula [the equation is stated, but unnecessary here][188] was first obtained by H.A. Lorentz, under the assumption of contraction of a rod in the direction of its movement in order to overcome the difficulties of the ether hypothesis, but his theoretical grounds were not satisfactory. On the other hand, Einstein started with the following two postulates: (i) Special principle of relativity: A physical law

---

[187] Published by the Mathematical Society of Japan, The MIT Press, Cambridge Massachusetts and London, England. Ed. Kiyosi Ito. Vol. II, p. 359 B---as already cited in the book.
[188] The Lorentz Transformation and The General Transformation of Co-ordinates are discussed by Professor Eddington in *The mathematical Theory of Relativity*-- Sections 5 & 15. In any case, the Japanese mathematicians did not think much of the Lorentz Transformation, neither did I, and it is not strictly relevant here either.

should be expressed in the same form in all inertial systems namely, in all coordinate systems that move relative to each other with uniform velocity. (ii) Principle of invariance of the speed of light: The speed of light in a vacuum is the same in all inertial systems and in all directions, irrespective of the motion of the light source. From these assumptions Einstein derived [the Lorentz Transformation] as the transformation formula between inertial systems x = (ct, x, y, z) and x_1 = (ct_1, x_1, y_1, z_1) that move relative to each other with uniform velocity v along the common x-axis. This was the first step in special relativity, and along this line of thought, Einstein solved successively the problems of the Lorentz-Fitzgerald contraction, the dilation of time as a measure of moving clocks, the aberration of light, the Doppler Effect, and Fresnel's dragging coefficient." The time Einstein used for all this, according to him, was the Lorentz local time, provided, it can be defined as 'time, pure and simple'---meaning it is all there is of time; this qualification is very important. The Minkowski theory of four-dimensional continuum came later.

Apparently, time can be so defined, for it did not hinder his work until Minkowski intervened.[189] It means every time is somebody's local time. To overcome that we have learnt to mechanise our time in the clock for general use---but, and this is the crucial point, it is based on the earth's motions. Therefore, earth time as derived from the earth's regular motions is all the time we have or can have. This is definitely the view from Einstein; and that was the end of the matter rationally, as Einstein would have it.

Yet, by using concepts like the 'homogeneous Lorentz group', 'time reversal' 'space reflection', 'parity transformation', 'the

---

[189] Through all this time was and still is the same. The second is the same because it is based on the earth-year, year after year after year. With Minkowski or without him time is the same. This is what alerted me to investigate what time is and found that we can never know what it is. We use the same method and therefore get the same results---units of passing time, or properly units of passing cycles as the rates of passing time. Ten orbits of the sun are ten years, yet they are mere physical orbits. By mathematics the units can be manipulated; so mathematicians can use the Minkowski formula and manipulate time units to remain the same as we are used to, and say it works and therefore must be true---except that the imaginary time coordinates give the game away.

proper Lorentz group', and so forth, none of which made the original Lorentz formula satisfactory, as the Japanese mathematicians aver, cosmologists, the interpreters of general relativity and pure mathematicians are propounding impossible theories about time and call them Einstenian, but mostly inspired by the Minkowski theory that Russell and Eddington described as arbitrary and fictitious. I hope I have made myself clear to avoid any misunderstanding. The mathematics for these concepts are so daunting that British readers, who often confess they do not understand mathematics, are alienated---but that is wrong and not very helpful as far as I am concerned since I write in English.

Intellectually Mathematics is for demonstrations; it is not good enough for original discoveries. The discoveries come from conceptualizations or philosophical thought. The valid logical concepts (or the logical train of thought) come first; that is what really matters when you are promoting new idea. So if any theory can be understood without its mathematical demonstrations there is no need to alienate readers who are not happy (to put it charitably) with mathematics. For this reason I rarely state mathematical equations in full, the reason for sticking to '$s=ct...$' throughout this book.

By the Minkowski theory, time travel is said to be possible. I am afraid that is not true. But of course unless the underlying physics is true or correct the theory cannot be physically effective or true as regards our physical world. All notions of time travel are sheer humbug, because Minkowski was wrong. His formula was based on arbitrary ideas and therefore logically invalid. I have to add that it is quite unacceptable to try to conceal logical errors in thought with mathematics as Minkowski has done. Einstein is not to blame; he was literally coerced by mathematicians to accept the Minkowski formula, but either way relativity (special or general) is not affected.

There is credible evidence to justify the claim that relativity is not affected whether time is regarded as the same as space or independent of space. Even I should say that the evidence is not only credible but also strictly logical. In the special theory of relativity, Einstein made time independent of space. Why didn't he go back to make the time the same things as space with the 4-D geometry after he adopted the Minkowski formula? To me the reason is this: good old Einstein was no fool. He despised the 'superfluous learnedness' of the mathematicians who were tampering with his theory; yet he needed the support of the scientific public, the majority of whom were coercing him to adopt the Minkowski theory. At first relativity was universally ignored; so he acquiesced; but he was no fool. I believe he knew that either way relativity is not affected. Thus he left special relativity as he originally conceived it---where space is separate from time, and we still live in a special relativity world.

As noted earlier in some of the notes, whether time is the same as space or separate from space, the second is always the same---time is always the same; and Einstein wouldn't miss that point even without his brains. The difference between the two versions of time is philosophical, and it is this: thanks to Einstein we now know that every time is somebody's time; that there is no one (overriding) system of time that covers the whole universe. A second here is not like any other second anywhere else. The Lorentz concept of local time was called 'time, pure and simple' by Einstein, meaning that it is all the time there is, or can be had.

Alternatively it means that is the only way to get any time at all. Therefore anybody anywhere can get time, generate or create time, thus breaking the mould of Newtonian absolute time and virtually setting in motion the beginning of a new world not guided by cosmic time---but what is the nature of our time and where does it come from? We still have to answer this question and that is what I am trying to do in my books.

Considering how fundamental time is in human affairs, this is a philosophical concept of a revolutionary kind---there are as many times as there are inertial bodies. Russell put it most succinctly as already mentioned: "...There is no longer a universal time which can be applied to any part of the universe without ambiguity" In his judgement this was 'perhaps' the most important of Einstein's intellectual contributions, and I keep adding that there is no perhaps about it. It was the most important of anybody's philosophical logical scientific, mathematical, cosmological and astronomical contribution to human knowledge. It means our time cannot be applied anywhere else. He's right. So far we know only of the time we have created as our local time to suit the earth's motions. Yet time goes to the roots of all existence. Anybody anywhere needs time for every activity (it is the waiting period in all activities.) Everything we have ever used for time amounts to quantifying it without knowing what it is. I believe this will apply to every 'Being' in any part of the universe. We can only use repetitive cycles (and count them like the years) to show how time is passing only, not what it is.

Time is closely associated with 'Being' or 'Existence'. There is a natural aspect of time in our conceptions. Professor Eddington called it 'the internal time-sense', being the sense of duration. But we also know that it requires points; therefore it cannot be the same as 'Being'. For it could not have been invented without using points; inevitably, sentience was required.

Thus we must look for an aspect of life that can be mathematically linked to repetitive external cycles (the years, for example) to yield units of time that accord with physical reality (the most obvious is Day and Night), and can also be mechanised in a clock, which, once achieved, should be seen as man's greatest intellectual (scientific, mathematical and philosophical) creation or invention---and Einstein led the way to the most logical theory ever; that is the reason I maintain that time was Einstein's greatest achievement, for that is why we can logically link physics to philosophy by means of time, as we can also link them by means of the quantum theory.

In post relativity physics time must be based on our roots, and I think it is now seen as such. But how? The mechanics must be explained. Well, time is now logically conceived as something whose roots in our minds are based on the sense of duration (or the capacity to experience duration as an internal sense of time, as Professor Eddington put it: that is, of the impressions, images, and so forth, of things enduring, since it takes time to endure or linger), and therefore part of the physical and physiological mechanism for memory---if memory is defined as 'the capacity to repeat'.[190] And that is one way of linking physics to philosophy.

---

[190] This goes to the roots of our existence because it is part of the mechanism by which we gain knowledge and remember it---which is the sum of the contents of the human mind; part of that knowledge by which we live is the concept of time, that is, of things lingering. It takes time to linger; added to the repetitive external cycles, we get time units to mechanise into clocks.

The Minkowski formula contradicts this supposition; I concede that his theory would be the greatest achievement of the human mind (linking physics to time via space); but then he based his theory on imaginary time co-ordinates, and therefore it is regarded as logically flawed. So it means the physical world or the world of sense is not four-dimensional; any theory that assumes that it is will be contrary to the known laws of physics. I have called the Minkowski formula an attempt to deceive nature. I am not going to accuse him of a more serious intellectual crime than that.

I confess that it is (or was) technically difficult but I have somehow managed to show why the Minkowski formula seems to work. This is just an elaboration of the gist of the last few paragraphs. But for your (and everybody's) benefit it will do no harm to repeat it. The question is whether time is secular and originates on this planet, or it is generally in the universe to be invoked with the appropriate mathematics (and mathematicians seem to prefer the cosmic interpretation that implies the existence of God); secondly, whether i can invoke such a time. It cannot because imaginary time does not exist anywhere except in a dream. However, time (or the seconds) is always the same in our minds whether the time is seen as separate from space or part of 4-D geometry; and that is what Minkowski exploited. The seconds are always the same in the mind, but how do they get there? The answer is by using repetitive cycles to quantify time, and, I have shown the technical methods in my works as a link between the sense of duration and external cycles---the years, for example---and Professor Eddington said much the same thing. He wrote: "The rough measure of duration made by the internal time-sense are of little use for scientific purposes, and physics is accustomed to base time-reckoning on more precise external mechanisms." (ibid, Ch. 1.8---p23.) So there are two aspects of time: the physical or practical and the psychological or the sense of duration.

Here is a brief explanation of the idea that Duration x external cycles = time in units, for usable time is in units only, known as 'quantified time': the year is just one unit of time, and all other known units of time are derived from the year with points or mathematics, thus making them also discrete, including the cesium units, since they have to be related to the second to make sense. Let us assume there are no clocks: now suppose you see an image on TV, then it goes away after a while. How long it was there is its "duration" in sense or in the mind, otherwise known as 'the internal sense of time', or 'the internal time-sense'.

It must be clearly understood that duration implies the passage of time---i.e. during the period (or the life) of an event. But obviously it is not enough; it is not the time you can mechanised in a clock for general application.[191] Something else must be added to the duration, namely, it must be converted to units of time. This is easy to do, for we know how we get the earth-year as a unit of time---round the sun as a cycle. In fact, the year is our basic unit of time, metaphysically. To have more units (or years) we go round the sun again, otherwise there are no naturally occurring time units, or years. Strictly speaking, this is not a theory. The process of creating time units to mechanise into clocks is the metaphysical origin of time as a union between duration and how it is broken down into units---or cycles, or the process of quantification.

To convert duration to time units, you will have to use repetitive external cycles and count them (like the years) as the numerical units of time for the duration of whatever event or events are involved. But this has been done already for all of us, hence the clock.[192] This procedure will be the same for any sentient beings anywhere else in the universe; we can only use repetitive cycles to create time in units. That is the only logical way to obtain time in numerical units---otherwise time is 'silent ageing', 'silent motion', or 'the silent passage of existence' in a perspective, all of which are useless to science and logical thought.[193] Without mathematics, logical thought (in abstraction or in any depth) is not feasible, and you need to apply mathematics to duration to get time in numerical units suitable

---

[191] For any purpose where time is to be cited (as the additional co-ordinate of relativity, for instance), you need to have the time in numerical units, which can only be achieved with external cycles.

[192] This is easy to understand. You can even tap your finger, and say, for instance, the duration (or the life of the relevant event) was for so many taps of the finger.

[193] So our time units in the clock are not related to time at all---nobody knows what it is. The clock only gives us an idea of how much of time is passing by. Therefore the notion of measuring time by the clock is totally wrong.

for logical thought. In sleep or coma, time will be passing by. However, when you come to and want to know the time or how long you have been senseless, you will need mathematics based on some kind of repetitive motions or cycles to be able to have the time in numerical units.

We on earth here use the earth-year as sub-divided down to the seconds, or the cesium units, to determine the time "during which" our putative TV image was there---that is endured or lingered. The term 'during which' means duration, but it is not enough as time.[194] You will have to relate it to some of the sub-units of the earth-year to get the appropriate time. (You will say the event was 'so-and-so' long; that so-and-so length is obtained elsewhere and applied to the event. In all cases we have to use repetitive or regular cycles. Metaphysically there is no other way of obtaining units of time to work with as the rates of passing time---or of time for short.[195]) Thus we apply some of the earth's sub-cycles to 'duration' to get the time for it---to get the time 'during which' anything was there; and that means converting duration to time by linking it to external cycles. The external cycles themselves do not constitute time, either. They are given durations, or periods of mental lengths (during which they were there--- in the mind), before they can constitute time: a month is longer than a week; a. second is shorter than an hour.

It is by the sense of duration (during the life of an event, an image or an impression, as one experiences it) that one can determine which unit of the external cycle to apply to it to get the time, *or the number of cycles during which it was there*. The two statements (time and number of cycles) are exactly equivalent.[196] So that you can say it was there for one cycle=one year, or a shorter period. Alternatively, you can apply any of the sub-units of the year's cycle to the event. Due

---

[194] Because images, impressions, events and so forth, can linger in the mind, they are obviously connected to the mechanism for memory, which is defined in science as "the capacity to repeat".
[195] That is why 'time' and 'the application of time' are two distinct operations of the mind, but often they are conflated leading to unnecessary mysteries about time, as discussed earlier.
[196] For example, the numbers of the earth's cycles or orbits round the sun are known as years, or periods of time.

to our use of clocks, we have forgotten that this is how we created and mechanised time for the clock/Calendar system.

The metaphysical question is whether there is any other time. Well, the passage of existence is regarded as time, but it is not quantified or usable time. My own opinion is that the method described is the only logical system that any sentient beings in the universe will use to get their quantified time because logical thought must definitely be universally the same everywhere as we have it on earth---for it is a process of reasoning about percepts. Nobody can live rationally in any part of the universe unless he or she adopts reasoning according to percepts, and that is what we call logical thought. It may get more complex and mathematical with increasing volumes (and in non-demonstrative inferences), but ultimately thought must be based on percepts, or material reality. And material reality for any sentient beings is bound to be broadly similar to what we experience on this planet or in our Milky Way.

Even Idealism, before it was successfully refuted by G.E. Moore, was somehow related to percepts: if you cannot see anything the question as to whether it is mental or physical will not arise---simply because the 'it' will not be known. If the Irish philosopher, Bishop Berkeley, ever saw anything (say, the pen he wrote with, the paper he wrote on, the table and chair in his study), then there is no argument. In any case, Berkeley rather confirmed the existence of the quanta hundreds of years ahead, without knowing it, as Bertrand Russell has pointed out in his History of Western Philosophy.[197] The most interesting refutation of Idealism, of course, is the quip that a train at a station that can be seen to have wheels cannot be said to lose the wheels when in motion just because the passengers are not seeing them, or looking at them, as the Idealist ideology requires.

About myself (since you asked) and reactions to your criticisms, I confess I felt a little sad, not annoyed but rather sad, that you should mention the Minkowski mathematics to me. At this level it is most unfair to assume that I could be ignorant of the Minkowski mathematics. The truth is that mathematicians have overlooked the caveat of Professor Eddington (in his monumental opus, The Mathematical Theory

---

[197] See his essay on Berkeley in that book.

of Relativity), to insist that Minkowski has changed the nature of time with his formula. To repeat: he said plainly that the Minkowski formula would be ideal for describing phenomena, but, while mathematicians may consider it as useful, they must not forget that it is arbitrary and fictitious (Mathematical Theory of Relativity, Ch.1.1.)

Yet what is the situation now? We find that what Professor Eddington and Bertrand Russell have both described as arbitrary and fictitious is making mathematicians shameful because they allow their work to be guided by it, and out of which such mythologies as time travel become scientifically possible. As always with poor old mankind, people show how clever they are as books about such subjects sell millions. I am not surprised, but if Minkowski was right why didn't Einstein go back to amend special relativity?[198] I think he was coerced to use the 4-D geometry in general relativity to make it easier to understand. As a result general relativity is now in a hopeless mess. Yet again, either way, the basic postulate of general relativity (the curvature of space for gravity all the way to inferences about black holes) is not affected. Einstein was no fool! Footnotes of his theories will come to replace the footnotes of Plato's theories in philosophy. That is my prophecy. Einstein must have known that, intellectually, he's something like God.

Finally, I have already indicated that we can link physics to philosophy through the relativity concept of time; we can do the same thing, again, through the same Einstein's concept of the quantum, or his "Light Quanta" theory, too. Originally he called it 'a hypothesis'. But it is no longer a hypothesis, not even a theory, as scientists have confirmed it as a fact of nature

---

[198] Another question is when cosmologists and mathematicians are going to realise that earth time is not applicable to any other world, frame or metric, outside this planet. All the work they have been doing in general relativity since Einstein is vitiated because Minkowski is wrong and our time is not applicable to the metric of general relativity---it is a different frame, or world, 'pure and simple' as Einstein would put it. We are subjecting the entire cosmos to earth time; yet even dull school boys know that the cosmos is different and complex; besides out time is based on the year. But the year is indefinable as well as being just one orbit of our tiny sun; it, is bound to be an inaccurate temporal yardstick to apply to the cosmos. Ten years are ten orbits, but how long is that in cosmic terms?

through QED.[199] So the man I call the only true God we know did a lot to justify his divine appellation, a hundred times over.[200]

The quanta are seen as light. A single one will be a solitary speck of light of a specific hue; and as small as they are, scientists have invented a machine for counting them one by one (QED is said to be the most well-established of scientific theories, and it's all concerned with quantum interactions). En masse we call them quanta---or the smallest pieces of matter that can exist. We can link physics to philosophy through the quantum because it is light, and we have always assumed that we see only by means of light. That idea is rather a misconception. In actual fact, we never see things at all; we see only their images.[201] Nobody has ever seen anything but its lights only. Physiologically it is quite impossible to perceive anything. Seeing occurs in the brain not on the eye. How, for instance, can we fit a house physically into the brain's tissues? Rather we think we see a thing because its lights consisting of these very small quanta reach the eye with its exact image, and thence to the visual cortex in the brain.

So we see only the surface lights of things, not the things themselves. To quote Russell on this again, another philosopher conceived his theory of vision just to account for this very fact. It is precisely what the Irish philosopher, Bishop George Berkeley, stated in his philosophy, although he interpreted it wrongly. "Berkeley advances valid arguments in favour of a certain important conclusion, though not quite in favour of the conclusion that he thinks he is proving. He thinks he is proving that all reality is mental; what he is proving is that we perceive qualities, not things, and that qualities are relative to the percipient." This is permanent proof that we do not see things; we see only their surface qualities in the form of their exact images as constituted by their light emissions, or radiations, or by means of quanta---the smallest bits of matter

---

[199] I have my doubts because the quantum, as energy-second, is time-dependent. The energy is natural; the time is our own peculiar time. Can we apply it to the rest of the cosmos without ambiguity?

[200] May I remind the reader that this chapter's style is due to the fact it was originally addressed as a letter to some readers on the internet. It is included here in case anything I wrote then could in some way assist the reader's understanding.

[201] Plato conceived his Theory of Ideas to account for this fact.

in existence---and which is also light or what we know and call light except that in nature there is no such thing at all---light is transmission of matter from one body to another.

We interpret this as meaning that we see only the smallest parts of matter that, by their nature, can never stand still and always radiating about through their interactions with the electrons of other matter. The implication is that we see matter, and yet do not see matter. We see only matter's smallest parts flying about---but they are also matter.[202] The philosophical importance is that pieces of matter do actually fly off things with their exact images; when we capture some of them on the eye we see them, with all their colours because the quanta are naturally coloured. There are intricate technicalities in all this; physicists will not put it so bluntly or crudely.

They will have to add a number of fine qualifications. But to philosophers this is enough to work with, for it so important it makes the Platonic theory no longer valid in so far as the external world is concerned. Capturing quanta from things to see them means, as Bishop Berkeley supposed, seeing the surface qualities of things and not the things themselves---the main reason is that all seeing is 'Tele-Vision'. When you are very close to a thing it is difficult to see it properly; if it is too close to the eye you might not be able to see it at all without confusing blurs. Seeing by means of quanta is precisely like photography. It has to be at an appropriate distance for a clear image to come through. The only problem is size. But, frankly, we have been on this planet long enough; it's about time we got used to our inability to understand or explain everything. However I have my own theory about size in vision.

Size presents a special problem that can only be settled with a dose of speculation as opposed to solid facts. But the speculation is based on the facts given above, and they have

---

[202] Because of QED, we cannot (at least, I cannot) escape the fact that the quanta are the smallest pieces of matter out of which all other forms of matter have emerged through coalescence. The Scientific Industrial Complex has found another way of spending our taxes, so scientists can go on (who can stop them?) building larger and larger 'Atom Colliders' to split atoms in search of the basic building blocks of matter; but for me, my notions of the origin of all forms of matter begin and end with the quanta, the particles of light, thus making light 'What Is' in logic.

been proved more than sufficiently. All that is required is that the inferences based on them should be logically valid.

Now, because it involves the quanta, seeing is obviously like the electronic scanning of (or in) computers.[203] We imagine that human vision occurs atom by atom, since the quanta are the smallest sub-atomic particles---they interact with matter through their sub-atomic parts, particularly through the electrons of atoms of matter as is well-established in QED.

This well-known electronic process gives a clue as to how size is handled by the brain. The thousands of pages in a computer file are not spread physically in the computer as we spread documents on a table. Similarly, even a single A-4 paper cannot fit into the brain's 'bloody' tissues. Let's face it, when surgeons cut open the brain they see only bloody tissues; yet these same bloody tissues give us awareness of non-bloody percepts; and we know that is possible only at the sub-atomic level of physical reality. At this point a certain amount of speculation creeps into our thoughts.

We believe that seeing the paper (of any colour) involves electronic processing of quanta in the brain, where the bloody tissues our surgeons see are not bloody at all but part of the electronic processing of the quanta of things in people's heads. The paper's size is electronically scanned end to end. Since it has its own colour, when it ends, its colour will also end---and another thing's colour will take over. Let us suppose that the paper is white. At every end of it (all four corners of it) different things' colours will take over. We call the four white corners 'the paper'. The other colours would belong to something else, because the paper has ended, and something else must be there not void. Even void has its own colour to identify its presence. This is how one thing is one thing, and another thing is also another thing in vision. Needless to say, given the nature of human motives, confusion in vision can be induced

---

[203] I have to sound a warning that, although the computer can give relevant examples regarding the operations of the human brain, all such examples are copied from the brain. Let me explain this contradiction in terms. Scientists take suppositions from many sources and test them; some work, others do not; but even those that work are poor imitations (and certainly less complex) of how the brain actually works, which is not available for scrutiny.

through this method of visual perception. The white sheet of paper will stand out; when it ends, its colour will also end to show that it is no longer there before the eyes. The end of an image is marked by change in colour. As far as the brain is concern, the process of seeing the white piece of paper is like the computer scanning it minutely atom by atom at the quantum level, which is very small indeed.

In this way, objects of any size can be visually perceived (electronically scanned), without having the things (of so many different sizes) physically lodged in the brain tissues. By this suggestion size in vision is determined by volume, shape or form, colour, space and position---all of which can be interpreted as 'different colour patterns'.[204] The light we have to shine on things to see them create billions of quantum points of emission, "scanning"[205] the objects of vision as if with a torch in the dark but from billions of points, thus conveying size in vision. So light does not illuminate objects for us to see them; rather they cause objects to emit their own lights from billions of

---

[204] There is ample support for this view because the best scientific definition of size in vision is called "pattern recognition"---sometimes the phrase is used to represent the actual process of vision itself. There is evidence in physics that the patterns are created and conveyed by quanta to the eye (this is what renders the Platonic theory false); and so long as the objects are there and the process goes on through the excitement of the quanta, the objects and scenes, however big, will be seen and stored and be capable of being recalled in memory or dreams--- exactly as they were or as messed up by the brain.

[205] Scanning is not the correct word; but there is no correct word. What is involved is so unique there are no analogies. You have to pity those who write about the quantum with our "crude" traditional means of communication. The nearest description is to say the individual quanta are so small that the total number required to give vision of the wide open sky would fit on to the sharp point of a needle with copious room to spare. In addition, they will persist, making it look like scanning. So long as the thing is there, the quanta radiating from it will persist; therefore, vision will occur----the eye is bombarded by the quanta from the thing. Remember that they are trillions x trillions x trillions; they are persisting; they are radiating from all aspects of the thing; thus they will continue to give vision of the thing. The visual process is better described as the absorption of quanta, the quanta from things---we do not see the images of objects such as is supposed on the old Platonic theory. We absorb their material signals direct from them; if they are big, the visual process is like scanning their surfaces---not quite, but almost. Their full sizes will be seen so long as they are there because every atom in them will be emitting photons to reach the eye of the observer.

quantum points on the thing's surface that carry their exact images.[206] Seeing in this sense makes the Platonic Theory of Idea redundant. Let me repeat this because the Platonists are pretty stubborn: a light source does not illuminate the objects we see. We need the light all right, but it does not illuminate objects. By the miracle of quantum computation and interactions, this is not at all surprising. In fact, there is no such thing as illumination in the universe. According to Niels Bohr Light is "Transmission of energy between material bodies at a distance". Thus we should regard the quanta from the light source as interacting with the atoms of the things we see, rather than illuminating them. Emission and absorption of radiation is corpuscular. Since in the quantum world we live in there can be no illumination in the universe, Plato was plainly wrong. As always, the problem is to get mankind to forget about Plato, the pioneer, and look at the new world now presented by the quantum theory---which will be difficult because Plato can be used to justify religion.

The inevitable conclusion is that both science and philosophy (or physics and philosophy) have identified one entity as the ultimate cause of the nature of physical reality, or "What Is", namely: we see the world through the quanta; also in all its physical analysis of the nature of the external world, physics has identified the quanta through QED as the cause of all physical reality in its multitudinous forms. So physics and philosophy are linked in the existence of the quanta, another of Einstein's discoveries. He has therefore done more than we expected of God---even Plato is no longer interesting.

On the other hand, if we are to replace the Platonic Theory of Ideas with the new concept of "Quantum Signals" issuing from things to enable us to perceive them visually, then we have to adopt the corollary, which may be called "Coded Signals In Perception", meaning that things are given codes in the brain

---

[206] I hope the reader understands what I mean by 'quantum points'. A point so small that billions of them make up a pencil point on paper. Thus the image of anything being scanned can be seen by the human eye---the whole size does not have to come to the eye only their quanta will do so to trigger a chemical process in the brain so that when arouse or invoked the same images can be seen again and even in many guises. I cannot think of any other logical manner in which seeing by means of quanta can be explained, especially in dreams or the mind's eye.

for purposes of memory (and all cognitive processes) and not the things themselves physically, since all bulky objects, bigger than the quantum are just too large to fit into the brain as they are seen externally. Hence the theory that they are perceived by way of scanning by quanta (the particles of light), and stored chemically seems logical.

This sounds confusing, so I will do my best to explain it well. I think electronically coded signals cause vision, and that it is the reason we can dream of objects and of events with the eyes closed. REM seems to indicate selectivity and scanning as in the computer.[207] It is interesting that scientists have found that vivid dreams are particularly associated with REM.[208] The electronic coding may have internal and external aspects. Internally, it is obviously part of the mechanism for memory. Externally, the brain must have created categories of 'perceptive images' (images of classes and groups of objects) over several centuries: the figure of a person, the flight of birds, quadruped motions of animals, billions of shapes and forms of objects, and so forth---a long list of categories, literally infinite. But their coded signals are there otherwise they couldn't be seen or dreamt about when they are not physically present. Anything new will get its own category; anything known will have its established category already in place; thereafter associated objects will be added to it. What is on two feet would be assigned to the category of two-footed creatures, including people; what is flying, to the category of flying objects, and so on. We imagine that when something is apprehended, the brain is able instantly to place it in its proper category and recognise it then and thereafter.

We have to remember that neurons perform several billions of calculations a second. No computer can ever come close. That is how complex the human brain happens to be. Internally, the codes of things come into play when the thing is invoked (through the appropriate stimuli); so that however big, it can be

---

[207] A Nobel Prize has even been awarded for this theory of memory. The American psychiatrist, Professor Eric Richard Kandel, was awarded the Nobel Prize in 2000 for proving that memories are caused by changes within neurons and synapses. So this theory of memory and how we store ideas, images and events, large structures and even the wide open sky in our mind's tiny, bloody tissues has scientific backing.

[208] By the Sleep Research Centre at Loughborough University, for instance.

seen in the mind's eye, because the coding system is electronic on the quantum level. These are all guesses or speculation, but without them vivid dreams cannot be explained without mysticism.[209] We are trying to imagine how the brain works; nobody has any cast-iron proofs.[210] However, what we have found so far sounds credible. The computer can help, since it works through the system known as 'pattern recognition'; and, as I have said above, in the matter of deciding how size appears in vision, we assume that the quantum signals from things behave as if they are scanning the objects from end to end. So the size of any object will extend to the end of the thing's image/colour or colours. **In vision colour is the distinguishing mark of individuality, and space is that of time.**[211] In this way it is believed that size is determined by shape or form, position and colour---all of which can be abstracted to 'different colour patterns', because even 'the void' has to have its own colour---its colour has to be different from those around it. A large object is seen not because it enters the head with its size but because it is scanned from end to end. It is the scanning process we see not the actual size. And this happens at the quantum level. So anything of whatever size can be seen.

There are aspects of the brain physiologically established from scientific medicine, computers, and human behaviours that suggest that the above theory may be close to the truth of how the brain actually works. Nevertheless, even if the suggestion is

---

[209] Mysticism encourages religion. Religion causes wars.

[210] However, as his final theory of the human mind, Bertrand Russell said that when a physiologist is examining the brains of a patient, what the physiologist is seeing is in his own brain---see the Chapter, My Present View of the World, in his book, *My Philosophical Development*. This theory of the mind implies a perceptive process that relies on signal codes in visual perception: we imagine that the patient's brain parts invoke the appropriate codes in the physiologist's brain to make him see them---as he would do in dreams with his eyes closed.

[211] To know reality (or what the external world is), you have to use colour and space; but the truth is, ultimately it all boils down to colour because the space is known through its hue; and these are unavoidable---because they are there and impose themselves on all and sundry---and at all times, unavoidably. This is tautology; but that is what reality is and how it is apprehended by man. Colour for individuality and space for time, together constitute physical reality or the existence of an object, or 'Being'.

proved scientifically, it will take centuries for people to digest such difficult ideas.

One of the difficulties comes from the fact that, if the quantum is a product of our time, as erg-sec, and the units of our time that give rise to the quantum are also products of the human mind (through a link between the sense of duration and external cycles), then the nature of human life ought to be re-defined because it is not properly conceived either in science or philosophy. For the quantum which is the basis of all matter as far as man is concerned, may not be naturally in existence throughout the universe in the form it appears to us, but rather materialises through our unique way of moulding elements to suit our unique nature; therefore I think the 'Copenhagen Interpretation' can be adapted to mean that the strange behaviours of sub-atomic matter may be due to the strange manner in which the human mind is 'generated continually' and interacts (and interferes) with nature, namely, not constituted as a solid matter (or mass) but put together by fleeting and highly perishable impressions at the electronic level---and always growing, changing, interacting and communicating.

It seems nature does not see us as special, or as human beings, but mere neurological robots with no right to claims of superiority or any metaphysical pretensions; perhaps we are more efficient than other animals in nature---well, maybe that implies that we are somewhat superior in a way, but I doubt that that is metaphysically significant. Nobody can have the last word because nobody knows what the origin (or purpose) of life is.

The importance of what Einstein did, even without knowing it himself, is this: since the dawn of civilization, philosophers have been trying to interpret the world to know what it is made of, or how it is constituted. Out of their inquiries, scientists arose to claim that the philosophers have it all wrong. In addition, they started their own lines of enquiry, the chief part of which is the physical analysis of the external world, known for short as 'physics'. Einstein did not know that the philosophers and scientists have reached a stage in their enquiries where all their theories coincide in the discovery of the quantum---as the most credible candidate of What Is in both physics and philosophy. As far as he was concerned, he was doing physics, just writing his ideas about physical reality. And the miracle is that man does not even have to infer this new idea of reality (as caused by what we can see plainly) from any complex theoretical postulates from scientists or philosophers. We see the quantum plainly as light. They are now telling us that we live in a quantum universe. Well, we see the quantum as light so we see what creates the universe we live in. Physics and philosophy are thus linked. Of course, it has taken a long time to come to this conclusion, more than one hundred years in fact. Einstein himself did not know it---unaware of the significance of what he had achieved. That is because the work of interpreting his ideas to come to this conclusion was difficult.

The most important thing was the new secular theory of time. Until then time was so mysterious that it had literally become (as I prefer to put it), the last hiding place of God after Charles Darwin. Because of the mysteries of time I can imagine religious leaders smiling smugly at the increasingly secular theories coming out of physics. But when time is interpreted as something that originates from this planet (because there is no longer a universal time), and can be seen as a union between the sense of duration and external cycles, time is liberated from the religions; and it happens to be the most important aspect of life, second only to how the life itself came to be in the universe of inanimate matter, and without which no civilization (to sustain life) could have been possible. With this secular theory of time, added to the theory of quantum, we can now agree that man's science will die with him when the earth ceases to be habitable. Nature is far from uniform; the theories that work for us here on earth will die with us; so let us make the most of our good fortune.

With the above thoughts in mind, let me hurry to point out that the importance of linking physics to philosophy is to defeat the murderous peddlers of stupid religious myths from all the religions. Nobody is honest about religion; for nobody really believes in God; it is all lip service because everybody is afraid of death and misfortunes, and piety is everywhere imposed by force and trickery, lies and propaganda. Religion was not always as bad as we see it today; it seems without organised religion organised society would have been hard to create. But we could have created it nonetheless. Better than the internecine wars. The problem is how to replace it with science since it is no longer relevant---and persuade Africans in particular that the Bible is not literally true.

14

WHY EINSTEIN USED THE ARBITRARY MINKOWSKI FORMULA

The whole intellectual, scientific and philosophical world are eternally baffled as to why Einstein incorporated the Minkowski theory in the field equations of general relativity, since we are now told that he never even bothered to understand it[212]. My guess is that he could see that it would not vitiate his theory and yet would win him support from the powerful mathematical lobby.

First, before we come to Einstein's controversial use of the 'arbitrary' Minkowski formula in general relativity, why should Professor Eddington also insist that 'we shall continue to employ it', knowing that the theory is fictitious? And he said so in the same sentence, too. It seems part of the reason is that mathematicians were insisting that incorporating time into space makes relativity easy to understand. According to the British science writer, Dr John Gribbin, "Minkowski's

---

[212] I do not believe that he did not understand it. When a man like Einstein says something like that it is either a joke or telling us that the theory is nonsense.

geometrical description undoubtedly improved the clarity of the special theory and is still regarded as the best way to understand it." (New Scientist, 2nd Jan. 1993, p28.) When philosophers of science claim that relativity (even in science) is not yet properly understood, the following is part of the reason, namely: the very mathematical interpretation of relativity used for its understanding is itself arbitrary and fictitious.[213] If the Minkowski formula used for the interpretation of relativity is arbitrary and fictitious, then how can anybody convince philosophers that they understand relativity? What is happening is that the theory of relativity is obviously true and therefore it works; thus the many branches of science have had brave scholars to interpret it for them and the minions just follow their lead; but if you confront any of them to tell you the epistemological status of relativity, they either stall, dodge the question or invoke the Minkowski mathematics and stress that even Einstein praised him. Not many scientists have heard from David Hilbert that Einstein never actually accepted the Minkowski formula for equating space to time with his 4-D geometry. The Royal Society aristocrats would consider this suggestion madness in the extreme. You have to have special talents to live in Britain as a scholar.

I have a sneaking suspicion that the reason I am everywhere ignored is that physicists and the Royal Society just cannot imagine what will happen to physics once the new terms introduced as a result of the Minkowski theory are discarded, since they have sunk so deep into the subject. But I don't see how they can blame anybody for that, since it was the founder of astrophysics, the very big establishment pillar of science who wanted to disprove general relativity but ended up writing the definitive mathematical interpretation of it, who urged us never to overlook the fact that the Minkowski theory is arbitrary and fictitious---as I keep repeating, must repeat and will keep repeating to the disconsolation of The Royal Society.[214]

---

[213] See the essay, "Commentary on the Theory of Relativity" in the Microsoft publication, The World of Mathematics, Tempus, Microsoft Press, 1988, Vol. Two, P 1083.

[214] It is true that Professor Eddington never mentioned Minkowski by name in his denunciations; but in a book dealing with the mathematics of relativity, we know the dubious theorist he was referring to---certainly not Albert Einstein. To call Einstein's ideas arbitrary or fictitious would be a

Science has not progressed through the propagation of myths at its foundations in the various subjects; and yet we have been told that the Minkowski theory, although initially useful for the understanding of relativity, is nevertheless shallow in the serious matter of studying physical reality. Let me explain, again, what I mean in plain language: philosophers know that what is worrying mathematicians and physicists is that the 3+1 formula (preferred by Einstein) for representing physical reality is not, strictly speaking, objective or scientific, because man has to add the time, and yet this time has been declared 'non-cosmic' or 'non-providential' by Einstein, meaning that man has a hand in its creation as 'local time'---one's own inertial frame's time.

The implication is that we first have to create our time to suit our planet before applying it in the 3+1 formula for the determination of the nature of physical reality. There is no idea in science or philosophy more serious than this notion of time. It means man determines part of physical reality.

Mathematicians did not like that very much. Unfortunately, the Minkowski formula is not logically valid either, yet mathematicians and cosmologist still cling to it simply because it is camouflaged with intricate mathematics. Yet you cannot fool philosophers like Russell and Whitehead with such elementary techniques.

For Einstein, at the time, what mattered was the time. For there is always time. Even in our primitive ancestry there was time reckoning with shadows, marks and the moons; also with the sun and crosses on walls to represent days, weeks, months and years. The scientific study of time aims to trace or establish its metaphysical nature (since metaphysics implies or includes logical reasoning.) Hence, even as Einstein was debating time he was using time---he had a time system to work with---and it was always the same. He abolished cosmic time but he was using time---where did it come from? Units of time based on the earth-year are always the same. The argument was (and still is) about how they are obtained.

I know some scholars can argue that we worry too much about time since it is not doing anybody any harm and so we should

---

hanging offence, and he knew it.

ignore it and get on with life. Yet there are four main reasons why it is of the utmost importance to know the nature of time and whether or not it is joined to space. First, it is so strange that it excites our imaginations to make us want to scrutinise it and know what it is. Secondly it is a catalyst in everything, mostly in chemistry and all things are chemically produced, even the human body is virtually a chemical factory. Third, it is exploited by the religions and religion causes wars so that scholars want to know what it is in case some religions contradict it to their doom. And lastly it is so closely associated with life that it dominates everything we do; we encounter it everywhere without knowing what it is and whence it came---the human mind does not like that. It wants to know what it is.

The new idea from Hermann Minkowski is the merging of time and space into one entity to result in 'space-time'---as time, or the new concept of time. Because of space, this idea affects physics and the nature of physical reality, and so it ought not be wrong to distort physics. Therefore, if the experts have declared it to be arbitrary, then it is a serious problem in physics, and wrong to be perpetuated as part of the canons of physics. As an ordinary writer, I have a duty to point this out whether it will earn me honour or dishonour. So far, it has only been disregard or disdain, snobbery and scorn, nothing serious. Since I am not important ignoring me is a waste of time.

So, then, on this planet time is always the same because we use the earth year to reckon our time----thus making the fractions of the year what I call 'standard units of time'. Whether time is supposed to be coming from God or Satan, it is always the same---the second is always the same. We have something similar in Newtonian science. He said God created the elements and set them in motion. Well, whether 'He' did or not, the motions of the elements remain the same. Similarly, Einstein was not bothered, since he was known to disparage the mathematical interpretations of his work---and didn't try even to understand 4-D geometry, we are now told.

The real logical quandary in all this is that, technically and mathematically, metaphysically, cosmologically and all, the basic premise for equating space to time in the Minkowski formula is the square root of minus one, which, as every schoolboy knows, is used in mathematics to represent imaginary quantities or qualities. But how can we represent time (as real as we know it ticking away in a clock), with an imaginary quantity? Yet that is what the Minkowski formula for equating space to time is based on in logic as stated by Einstein himself, namely: "...we must replace the usual time co-ordinate t by an imaginary magnitude $\sqrt{-1}.ct$ proportional to it..."

This is what violates the crucial logical principle in human thought. So Eddington called it fictitious; Russell too, in his customary, good, logical rebuttal, stated: "...the philosopher cannot but feel dissatisfaction with the apparently arbitrary assumption about interval..." Little wonder that the magnificent and incomparable Einstein never believed in it---no matter what the mathematicians were saying, led, as we have seen, by one of the greatest mathematicians that has ever lived, David Hilbert.

Bertrand Russell said space-time is compounded in such a manner as to be convenient for the mathematician. That, obviously, is true; and that was precisely what Einstein took it for---and even used it to secure the support of mathematicians. It was not objective science; it was mathematical propaganda. Nobody should be surprised that Einstein used the theory and yet did not believe in its metaphysical credentials, for we study nature to reveal what it is, not as mathematicians want it to be. That crazy lot can even conclude that we should all be drown in the pacific on pure mathematical grounds, yet they create the mathematical rules for themselves.

Let us be clear about this. We are given a theory to the effect that a mathematical space, defined by space and time co-ordinates, gives us time ticking in a clock as 'space-time' and therefore space is four-dimensional including time. This is nothing but pure fantasy.[215] With his unique genius, Einstein could not miss the falsity in the formula even if he tried; so he must have known that metaphysically it could not be true of the world, or physical reality, or the reality of time in a clock. Of course, he used it, but only to help the mathematicians who said they could not understand relativity without it. It does not mean the falsity is washed away because Einstein used it in his field equations for general relativity. As quoted above, he used

---

[215] Rather 'relation between points', as Russell defined time, is much more capable of giving time ticking in a clock.

it mathematically to represent time---only in mathematics. Thereafter, time reverted to the standard units familiar to us in all other calculations---e.g. the seconds and hours and minutes did not change.[216]

---

[216] Even then, there is a deeper problem of time in general relativity so far not addressed, not even broached by any writer, thinker or mathematician. It is this: if general time permeating all through the cosmos (as cosmic time) is abandoned under relativity (as Bertrand Russell has stated), because every inertial frame is supposed to have its own time, so that Einstein could say "there are as many times as there are inertial fames", and since general relativity is not an extension of the earth's frame, how can we apply earth time's periodicities (as the earth's standard units of time) to the metric of general relativity? As far as I am concerned, there is no time in general relativity and earth time is not applicable there. It can be used there with mathematical adjustments, but as far as I am concerned, it couldn't truly reflect reality.

HOW THEORIES OF TIME CHANGED[217]

For the first time in human history, theories of time had to change because H. A. Lorentz discovered that it is changeable and therefore not absolutely fixed by divine authority.[218] For the mystery of time has always seem inexplicable and what human beings cannot explain they attribute to God, and claim they cannot define God himself because he has imposed restrictions on his identification! However by such myths we live whereas the life should not be possible at all.

Einstein said he borrowed his concept of time from H.A. Lorentz. He was the first to suggest that time might not be absolute or general such that a second here is a second everywhere else. He found that those with a moving clock read it normally; but others looking at it from outside the moving vehicle read it differently. How absolute time could vary in any

---

[217] This is it. Every writer on post-relativity time must answer this question.
[218] These short chapters merely serve to elaborate points already stated in previous chapters. I don't know if they will help the reader, but I have already committed so many literary crimes that another one won't matter too much!

circumstances, he could not tell. Therefore, he called it t^1, a mere mathematical curiosity. "He proposed to call t the general time and t^1 the local time. Although he did not say so explicitly, it is evident that to him there was, so to speak, only one true time t". (Abraham Pais, ibid, Ch. 7.)

But Einstein said he had an insight that all time might be like that---i.e. varying according to circumstances or situations. He wrote, and I repeat, "All that was needed was the insight that an auxiliary quantity introduced by H.A. Lorentz and denoted by him as 'local time' can be defined as 'time, pure and simple'." (Abraham Pais, ibid, Ch.7.) He then developed this into his theory of frames in the special theory of relativity: in a fragmented universe, it is impossible that a dynamic time would be the same everywhere. Lorentz also admitted that he failed to discover special relativity because he did not consider his discovery of local time as important. "The chief cause of my failure [in discovering special relativity] was my clinging to the idea that only the variable t can be considered as true time and that my local time t^1 must be regarded as no more than an auxiliary mathematical quantity" (Abraham Pais, ibid, Ch.8.)

Thence, all time became somebody's local time. Einstein even suggested that, "There are as many times as there are inertial frames. That is the gist of the June paper's kinematic sections". (Abraham Pais, ibid, Ch. 7.) Bertrand Russell concluded that, as a result, "There is no longer a universal time…", and asked further: "If cosmic time is abandoned, what is really measured by a clock?" (ABC of Relativity, Ch. 4&5.) Thus the origin and status of time were thrown wide open; and given the importance of time and how closely it is associated with life, it was no less a question about the essence of life itself that Einstein initiated. Some very clever thinkers like Gödel assumed that it means there is no time---yet we do have time (or something we call time) in daily use. The sad fact is that, instead of working on this problem, especially after Russell's query about what is measured by the clock, scientists have gone to sleep upon the cushion provided by the Minkowski formula, while knowing it is arbitrary, and then pretend that the issue is settled so that contrary suggestions are not even acknowledged.

From my point of view, the obvious inference is that we must define time as local time to conform to relativity. Since all time is known only in units, we can conclude that we get it as 'space-time', in the sense of creating the units of time (like the earth-year) with points out of the fabric of space---but not in the sense that space has been equated to time physically. Man can only get his units of time, like the year through the application of points to space. We need points, too, for subdividing the year down to its familiar fractions.

If we define time as suggested above, then earth time becomes necessarily discrete---exactly as we find in practical time: the year is only one unit of time. To get more years we orbit the sun again and again. Also, all the other units of time in use are derived as fractions of the year. An alternative name for 'discrete time' is 'Quantified Time'. For we have quantities of time, being the earth-year and its fractions, even without knowing the true nature of time. We quantify time by the use of repetitive cycles. Intellectually, this must be the greatest mystery on earth.

So earth time is discrete time. The problem is that discrete time has momentous implications that render most of our notions and legends of time false. But, first, it means the most important problem, which is the passage of time, is conclusively settled in logic, because it implies that we can never know the true nature of time, just how it is passing through only---and, luckily, that is what we know in practice as the passing years. The passing years constitute passing time, yet there is only one year. It is repeated to make time seem to be passing by through the succession of the units---i.e. the years themselves increase in numbers to pass by, yet in reality the year (the basic unit of time) is a mere physical orbit of the sun.

Consequently, there can be no kind of mythical arrow or arrows of time, which no one could properly define anyway, to account for the passage of time. The procession of the units of time constitutes the passage of time.

Ergo, in practice, the passage of the years is the passage of time. We know how this happens, namely we replicate the single year over and over again. That is the passage of time. I am stressing this on purpose because the irony is that the year is not even time. It is merely the physical orbit of the sun. So it means we do not know the nature of time, only how it is 'supposed' to be passing by with the passing years: we count the orbits of the sun as years. That is all we have (or know) as time. This is just one of the strange consequences of discrete time. I have published ten books (or monographs) all dealing with time and many aspects of the myths and legends surrounding it. In the meantime, let us consider a few of them in the next chapter.

16

SOME FACTS, MYTHS AND LEGENDS OF TIME

Nobody can answer all the myths and legends of time. There is religion, science and philosophy (and ferocious arguments bothering on violence) in the things people say about time; but nobody has any evidence for the concept of time we individually subscribed to. The only physical evidence is the Lorentz discovery that it is changeable, added to the Einstein analysis of Order and Simultaneity to confirm the Lorentz discovery.

1. The year is not definable. The length of the year, as time, can never be logically defined, for it is only an indication of how time is passing not the time itself. Other units of time derived from the year can be defined by reference to the year; but the year on its own can never be defined in logic. It does not seem likely that man can ever answer the question, "How long is one year?" The implication is that we can never define time logically, for the basic unit out of which all other time units are derived is indefinable. But we can see time passing by—e.g. the years. By counting the years, we get our ages. What

worries me is how astronomers are trying to estimate the age of any events in the cosmos in years without even knowing how long is one year. Also, the earth was there and we were inhabiting it long before we learnt to count the orbits of the sun as years. When did that begin, and when, therefore, is any event in the cosmos assumed to have began---from when, and how do you decide that?

2. Time Dilation is the name some writers have given to the Lorentz discovery about time. It is sometimes called, "The dilation of time as a measure of moving clocks". At other times it is wrongly described as "A strange phenomenon about time predicted by Einstein's special theory of relativity, showing that time slows down with speed". This can only be described as 'scientific mysticism'. Like all subjects, science has its share of mystic musings, often difficult to challenge because they are called 'scientific' or camouflaged with mathematics. In fact, the Lorentz discovery involved no dilation of time at all. It appeared that a clock had dilated, but in truth, the clock remained the same, since those travelling with it noticed no difference in its performance; only those reading it from outside the moving vehicle noticed any difference. If anything at all, it is 'clock dilation', not 'Time Dilation'.

Earth time can only change (or dilate) if the motions of the earth are altered. I have concluded that the supreme importance of Time Dilation is that it led Einstein to discover his theory of frames; and the theory of frames shows how life is possible in the universe---i.e. where gravity is mild enough for manageable regularities to occur, life to evolve and flourish, and perspectives are generated for sentience and intelligence to come into play, thus humanity. On the other hand, in general relativity where gravity is very strong, there is nowhere anywhere for life to evolve and flourish. In a book of this nature, repetitions are unavoidable, for after all the aim is to explain a theory using any kind of device that will help the reader's understanding. So let me stress again that I think it was through this supposition which originated from the concept of local time discovered by Lorentz, that Albert Einstein changed our concept of physical reality in the universe. As Abraham Pais put it[219]: "There are as many times as there are inertial

---

[219] *Subtle is the Lord...* Ch.7.

fames.[220] That is the gist of the June paper's kinematic sections, which rank among the highest achievements of science, in content as well as in style. If only for enjoyment, these sections ought to be read by all scientists, whether or not they are familiar with relativity. It also seems to me that this kinematics, including the addition of velocity theorem, could and should be taught in high schools as the simplest example of the ways in which modern physics goes beyond everyday intuition." He also cites one of the best presentations of relativity as Max Born's Die Relativitatstheorie Einsteins. Springer, Berlin, 1921. Translated as Einstein's Theory of Relativity by H.L. Brose, Methuem, London, 1924. I recommend this book to any reader who still wants to know what Einstein means to the world of science and philosophy.

Einstein gave us a new science that shows the limits of technical philosophy as well: general relativity is not for life and civilizations; special relativity is the place for them---how did he know this since it is true of the entire universe of matter? It is philosophical and yet provable; that is one of the scary hallmarks of Einstenian solutions to scientific questions. Thus, we can see that, even though he did not spell it out, the postulates that govern or determine inertial frames are actually three: the velocity of light; the laws of physics identical for all; and 'local time'. Therefore, there is no universal time. As mysterious as time is, it does not even exist without sentience---and even then we can only tell how it is passing and never what it really is.[221]

Of the three postulates, the most important is time; but I do not really believe that Einstein knew what he had achieved with respect to time partly because it is considered as 'philosophy', and partly because of the influence of religion, since their leaders did not want time discussed too scientifically to deprive God of his last hiding place. But because of time, I personally think Einstein was the greatest thinker in human history. And yet, look at how casually and humbly he introduced it: that local time may be considered as time pure and simple. That is how the greatest intellectual revolution in human life (overall) was

---

[220] I often attribute this to Einstein, but of course the words come from Abraham Pais's book, although the concept is undoubtedly Einsteinian.
[221] This three-postulate thesis has only just occurred to me; perhaps the academics can work on it in depth.

introduced---most admirable. In one of his Voice-Recordings Bertrand Russell praised Einstein's humility: "Einstein, whom I knew very well, was extremely lovable man, absolutely devoid of the slightest pretensions. You might have met him in the train without realising he was distinguished.[222] The most lovable man I have ever met." As has been said many, many times before, "The men and women who push forward the frontiers of knowledge are often unassuming, unhurried people with quite minds".

3. The concept of 'Clock Paradox' is also said to show that time slows with speed and that it was predicted by Einstein. The true situation is the following, quoted from Abraham Pais's Biography of Einstein: "Einstein rather casually mentioned that if two synchronised clocks C_1 and C_2 are at the same initial position and if C_2 leaves A and moves along a closed orbit, then upon return to A, C_2 will run slow relative to C_1, as often observed since in the laboratory. He called this result a theorem and cannot be held responsible for the misnomer clock paradox, which is of a later vintage. Indeed, as Einstein himself noted later...'no contradiction in the foundation of the theory can be constructed from this result' since C_2 but not C_1 has experienced acceleration." (Abraham Pais, ibid, Ch.7.) So the clock paradox does not prove that speed slows time. It may be able to slow down the workings of a solitary clock, but even then one clock is not the whole of time *per se*.

4. Twin Paradox. How the mathematicians love this one. They say, because of Time Dilation and The Clock Paradox, if a twin travels round the cosmos he or she will return to find his twin sibling considerably aged than him or herself. Those who propound these theories use them to argue that, as Professor Yourgrau put it, 'Time Travel is a scientific possibility'. In fact, unless s=ct, time travel can never be feasible on this planet. The Twin Paradox is a myth because time has not been equated to space. Space and time were made separate in special relativity, and they remain so, as we also continue to live in a special relativity frame.

Due to relativity's destruction of the foundation of religion with its secular time, some people are trying to turn the clock back

---

[222] I am quoting from my fallible memory, of course. I got it from the Nonesuch Record Series more than fifty years ago!

to Pythagoras through scientific mysticism. Practically every human being would like to believe that death is not the end. Thus time travel, echoing the Pythagorean Transmigration of souls implying there would be another chance to come back to life, is very appealing. Yet it all depends on how we define time, and nobody can define it logically. As mentioned above, even the yearly cycle we rely on for all units of time as the fraction thereof cannot be logically defined. Instead we use discrete time; and discrete time is spent when the units are ended; or to put it another way, when the units of whatever cycle used as time are ended, that's the reason we do not have years in nature, but only one year, so that we have to start another year at regular intervals to gauge our accurate ages in years. Properly I should say 31st December; but in a book of this nature the writer must be careful in his choice of words as any little slip could bring an avalanche of criticism from eager critics yearning for an opportunity to lynch him by the nearest lamp post. There is always religion in discussions of culture and time; and religion means hatred---kill the heretic, defined as "just somebody we don't like". I remember how harshly C.P. Snow was criticised over a minor grammatical slip in his "Two Cultures" because he managed to annoy both scientists and the sharp-tongue literati, particularly the latter because they were the guilty party.

Now, so long as we use or recognise the year as one unit of time so that we count our ages by the number of orbits of the sun (more so since all other units are derived as fractions of the year), earth time is automatically discrete, meaning that it consists of separate, individual and unrelated units of time. That is precisely how Professor Whitehead also defined time: ("A time system is a sequence of non-interacting moments.")

Discrete time cannot go forward; it also cannot go backwards---simply because it consists of units that are spent at the end of their lives, tether, reach, extent or range, etc---that is why we have the seconds, hours, minutes, and ultimately the years as individual units of time that can only be repeated to continue and make time seem continuous.[223] This system of discrete time as opposed to a continuous one (imagined by some

---

[223] Without a system of continuous time travelling by time (time travel) forwards or backwards is pure fiction.

people) has come about because the whole idea of numerical units of time are based on the day-and-night features of the earth. All time started with people counting the days. Without numerical time life is still possible. We think sometime in our distant past human being had no idea of time except that they aged[224]---but then the day and night system led people to count the passing of the days as the passage of time until logical systems were invented.

4 (a) Nowadays we hear a lot about cosmology and astrophysics. Taken together, the two disciplines have usurped the duty of philosophers, fair enough. The philosophers, if Sir Karl Popper is to be believed (see my notes in the Introduction), are not talking much sense to make him proud to be a philosopher anyway. There are journals for "The Philosophy of Science". The problem is that no thinker can challenge a subject based on experimental discoveries as science is. If scientists tell us what it is they find in nature, how can any arm-chair thinker oppose them with the opinions of even Plato as writers like Professor William Kneal are doing? So the philosophers of science end up writing sci-fi or the same old arm-chair philosophies, based on logical inferences. They think knowing a little mathematics is a substitute for insight. In fact, mathematicians follow ingenious insights and cannot create anything on their own at all, unless they think (speculate) like physicists or philosophers.

As against this, one wonders why cosmologists and astrophysicists are always asking for larger and larger instruments, or laboratories, to look mainly at what is happening in space, stars and black holes. It appears to me that astrophysicists (in particular) are too fond of expensive projects for studying the stars and interstellar space. Each time somebody dreams he seems to come up with new projects costing billions that solve none of our immediate problems here on earth. I agree they could lead to successes here on earth, but the cosmos is so vast and complex that man will forever toil in vain unless we limit our efforts and researches there to only matters of interest and relevance to our planet. The cosmologists in particular are convinced that whatever they

---

[224] This is not difficult to prove. Day-old infants certainly have no sense of numerical time yet they live.

discover in the cosmos is helping mankind to decipher the nature of the universe. Unfortunately, they are woefully mistaken; for the logical basis of all their suppositions is the Minkowski 4-D geometry or four-dimensional space derived from imaginary time coordinates. In so far as there is no such thing as an imaginary time coordinate, no idea based on it will be true, therefore they are wasting their time and talents, our time and money, and deceiving us, themselves and everybody else.

4 (b) Having digressed already, let me give some idea of the difficulty of trying to equate space to time. Strictly speaking this is a digression, but quite relevant to the 4-D geometry debate; it will only make my offence more grievous, but not fatal. The mathematics is so complex they call it 'counterintuitive'. Thus if you complain that it is either illogical or you cannot understand it, you are mocked for your ignorance of counterintuitive mathematics. Yet man does not live by mathematics alone; the concepts that we form of objects are more important, and they come first. Mathematics is supposed to express these concepts more scientifically. Because of this it is often claimed that the world is mathematical. It most certainly is not, but mathematics is the best human technique for revealing the compatibilities and usefulness of the many objects we encounter in the world or the universe. However, mathematics alone cannot save theories built on concepts of four-dimensional space that is constructed artificially with imaginary entities for the creation of physically recognisable entity like time---using ghosts to create physical realities is just not logically tenable. I think it is not even worth considering as part of a serious debate in physics.

In our present (post relativity) theories of physical reality space is so difficult to define that even Einstein confessed, "It appears that there is no quality contained in our individual primitive sense-experience that may be designated as spatial. Rather, what is spatial appears to be a sort of order of the material objects in experience. The concept 'material object' must therefore be available if concepts concerning space are to be possible. It is the logically primary concept."[225] In other words,

---

[225] Einstein, in his famous encyclopaedia article in Encyclopaedia Britannica, 13th Ed. 1926/26, page 105. He was writing under the heading "Space-Time". It is interesting how he emphasised 'concepts'. I have always felt that, in the study of nature, concepts come first their mathematical

what you call space depends on what you are doing or talking about, otherwise every space is occupied either by you or me, dark matter or energy, atoms, material objects known as bulky matter and so forth. As a result we are told that space is not uniform but dynamic.

Now, taking all this into consideration, what therefore, conceptually, is time as we know it? How can we define it? And how is this time linked to space to constitute one entity---not by anybody's bemusing mathematics but in physical reality in the way people know time in culture, society or life generally---even by primitive man? How time has been known from the earliest times to the present day---beginning from counting the days one by one?

By critical examination, we find that we can only use regular or repetitive cycles existing in our space (however space is defined), to give us units of time. (Hence the Lorentz local time concept which inspired Einstein.) We merely count the cycles (the years for instance), as the rates of the passage of time. There is absolutely no logical definition of four-dimensional space to account for this time. Everybody in science (philosophy and mathematics, or in life generally) is hiding behind Einstein's image because he accepted the concept of four-dimensional space for his general relativity. Well, we have now been told that he did not. My own opinion is that Einstein accepted 4-D geometry just to please his mathematical critics, and that he did so knowing that time is the same whether it is supposed to be four-dimensional or not, and therefore it could not vitiate his theory of gravity as caused by the curvature of space.

4 (c) Perhaps this is the appropriate juncture to go back to the concept of 'measuring' time in order to illustrate how we acquire the sense of duration artificially.[226] This subject was

---

expressions only come second despite the arrogance of mathematicians in mocking those that they assume to be not as versed as themselves in mathematics. Yet unless the concepts come first and they are mathematically compatible, the mathematics cannot follow or be found useful in that particular field.

[226] It has been mentioned above in one of the essays, but in writing a book of this nature, repetitions are inevitable. Some ideas that are used to support one point elsewhere may be used again to support another point. One prays that in the end one would be able to deliver a text of sufficient clarity to aid the

given serious consideration by none other than our most recent great philosopher, Bertrand Russell. When he realised that under relativity there is no longer a universal time, he asked: "If cosmic time is abandoned, what is really measured by a clock?" Absolute time was supposed to be eternal and divine. I call it the last hiding place for God after Darwin, because churchmen were so convinced time's mystery was so deep and serious that nobody but a deity could have imposed it upon the universe---mainly because it was supposed to be absolute and fixed (the easiest way to think of time still accepted by most people), whereby one hour here is the same everywhere else. Nobody imagined that time could change in any way. Then, at first, H.A. Lorentz, not Einstein, found that time is changeable. Or that it could be different in different situations or places; yet even he, the discoverer himself, did not believe it was important. Like everybody else, he continued to believe in absolute time that was generally permeating the entire universe. The important proviso was that it never changed. It is always the same and everywhere the same, too. Einstein rubbed salt into our wounds. He extended the Lorentz discovery to mean that "There are as many times as there are inertial bodies."[227]

Furthermore, time becomes the product of space. To put it bluntly, every planet will have to create or invent its own time out of its space. But time is very closely associated with life so what is it? It sent a shock wave through all thinkers, churchmen, philosophers and his own fellow scientists. The point is, as Minkowski observed, it came from experimental physics, from the Lorentz discovery, and therefore practically unchallengeable.

When time was accepted as running through the entire universe and the same everywhere, the year, which has been known for eons as the source of all other units of time as fractions thereof, was regarded as our chief means for

---

reader's understanding---which is the important thing.

[227] Those who argue that there is no time under the Einstein system are partly right---until you invent your time you have no time. They should apply that to the metric of general relativity and the Minkowski 4-D geometry!
However, we have something we use as time for the regulation of all activities and without which normal life is not possible. The intellectual challenge is to discover what it is.

measuring our version of time from the cosmos. The idea was that one orbit of the sun, which is a year, measures for us a unit of time (the year) that we can break down to the seconds.

After Lorentz and relativity, Russell could ask his question about measuring time because it follows naturally. The use of language is always instructive. The phrase "passage of time" acknowledges the fact that we age 'over time'. That life moves on either by external time or in response to its internally allotted time. Yet man has never been able to define time. So what is it? And now that this time has been shown to be fickle and not at all divine or imposed from above and not the same all through the cosmos, Russell's question should have become the focus of all the speculations about time: if time is not cosmic, then where is it coming from? It is easy to see how I began to formulate my theory of time. We have seen how other highly intelligent thinkers believe that there is no time, especially if one has got to invent his own time for his inertial frame or world he lives in. My reaction is that, 'but of course there isn't.' There is no universal time; but having to invent our own time means we have to have time---how do we get it? My answer is what I have explained in this book. We do not get it in substance; we trace only how it is passing with regular motions. There is nothing else we can use. Yet this is a major metaphysical shift in human affairs. For it means we can only know how time is passing and need not worry our heads about inventing theories to account for the passage of time.

Normally, questions put by great philosophers are not only important; they bring whole new subjects up for serious study. Plato's Academy was the pioneer. Apart from science which can spawn new subjects as it progresses (from biology alone we now have several subjects down to biophysics and even astro-biology and virology), almost all subjects in the universities arose through philosophers' enquiries.

However, about time alone everybody is afraid. Russell's question has not yet drawn the usual responses simply because nobody dares to suggest how and why the clock works except that it works. Most of the time human beings like to go to the religions to settle their minds about many of the quandaries in life. But about time alone everybody is confused---I believe Einstein was also confused. Nobody can offer any logical definition of time. What I have found is that we do not even try; what we use as time is just how it is passing by, based on natural cycles. We call this chain of physical periods of various lengths 'time' and get on with life. All the mysteries of time have come from mathematicians nourishing their innate mystical streak, philosophers trying to refute the mathematicians or religious leaders interpreting reality as 'created' by God exactly at noon on AD 40004.

To go deep into this question it is first necessary to investigate how we get what we call time by the clock or anything else because even Einstein did not define time.[228] But that, exactly, is the unanswerable conundrum more mysterious than anything except life itself. It wasn't so bad before Einstein. As I have said, people used to go to the religions and forget about the rest---that is, forget about whether or not what the priests were saying was true. Many religious people were (or still are) sceptical about many things and we know that some priest are rascals, but we still hope that there is God and that going to church constitutes good behaviour to guarantee eternal bliss after death. Belief in God derives from many things, including settled answers to life's problems, life after death, the need for succour or success in life, business and family affairs. And in a world where time is existing naturally so that we could just manufacture clocks to measure it in the specific units required, it was easy to accept that time was existing naturally throughout the cosmos as well and we merely measured our version of it (something men had done for eons without encountering any problems.) Nobody could even attempt to answer questions about where it comes from. Well, it seems time still does exist like a natural entity to the ordinary man, but not so to the philosophers (and science, too, after Einstein and Lorentz.)

But what is it, and why---since its metaphysics have changed--- are the mechanics of the clock the same so that time is unchanged for both the theoretician and the man on the Clapham Omnibus? Remember, we are asking this question

---

[228] He showed that it is neither general nor absolute such that the units applied to all parts of the cosmos with equal validity; but he did not define it, and rather praised Minkowski for his new theory of time---which, we are now told, he never even understood it. It does not diminish my respect for Einstein but increases it. Nobody can define time and I think he knew that too.

after Einstein. This means we want to know rationally what time that is not cosmic is or consists of. And it is the established academic practice (in both science and philosophy) to resort to logic in answering such basic questions about any aspect of life. By logic we mean the sum total of human intellectual capacity---observation, language and conceptualization.

The logical analysis of what we call time (from earth-time to atomic time, et al), reveals that the clock is not manufactured to measure time originally for us---that would be creating it metaphysically, which would mean that the clock could be changing the units in accordance with changes in the motions of the earth's orbit of the sun on its own automatically.[229] That is not how the clock works. We know that is not how it works because we have to alter its units from time to time to make them agree with changes in the earth's orbits or motions. *The clock does not give us the units of time by itself; it merely reproduces units of time deliberately programmed into it by the manufacturer. Reproducing time is not the same thing as measuring time. We could not know this until the new notion that time is not cosmic, and that the units are obtained elsewhere--- i.e. from a breakdown of the year---before being installed in the clock for reproduction, quite different from an act of measuring the time units originally.*

"There is no longer a universal time..." is what we know now---definitely. But life has flourished on this planet through trial and error. It has led to many habits. One common habit is the sense of duration, or the ability for recognising the lengths of different periods. That the sense of durations is a habit is easily demonstrated. If a child is ignorant of time (meaning he or she has not yet learnt to tell the time), saying you, as the parent, is going away for ten years would mean nothing to the child. Also if you were to tell a loan shark to wait at the door five minutes for his money he would probably smile happily; but if you said ten hours he won't be that happy. This sense of duration is not in natural existence; infants do not display it. They have to learn it after knowing how to tell the time. We only come to know duration after we have understood time and its uses. It is

---

[229] This is the reason the 'Clock Paradox' is not relevant in the interpretation of time. How any clock works (or performs) is not a metaphysical question about the original creation of time, as Einstein has pointed out.

an acquired characteristic that has become a permanent feature of the adult mind. We acquired it relying on absolute and universal time where we did not have to worry about how the time came to be in existence.

Presently we are told that it is not absolute and that time is rather cleverly worked out artificially by human beings through the intellectual use of points, repetitive cycles, a theory of numbers and the ability to count, thus making sentience absolutely essential. Yet even in these secular times (with our secular time system) we need the sense of duration. Without the sense of duration how can anybody tell the time periods of the various time units---that a second is shorter than a minute? It is the mental trait that helps us to know the lengths of periods still now. That one hour is longer than one minute and so forth, has been permanently structured into the fabric of the mind.

This is a unique example of how the human mind's creativity assists us in making the world habitable. We are rejecting cosmic time, but keeping the sense of duration it helped us to create and without which we cannot properly understand time.[230] We are probably not conscious of the fact that we are keeping it as an artificial aid to the sense of time. It has become an instinct, or a habit---so naturally woven into the fabrics of the mind that we cannot avoid it---and certainly it is so useful that we can never do without it. I think it proves that the sense of time and time reckoning also evolved. One by one some of the strange and mysterious elements in human life that were previously attributed by primitive man to God are being deciphered as purely secular, or that they arose through evolution. Or you could say we have now evolved to a level of mental maturity where cosmic or divine time is no longer consistent with the facts being discovered by science and therefore intellectually unsatisfactory.

Thus it happened that through the error of assuming time to be in natural existence, permeating the whole cosmos and the same everywhere so that all we had to do was to 'measure' it in the units required, we have acquire the habit of telling the

---

[230] I believe that is the main reason Einstein could reject absolute time and still use the time units it has created. For these units of time are based on the earth year and do not change whether time is regarded as the fourth dimension of reality or not. I am sure the great man knew that.

lengths of periods by the sense of duration built up in the mind. It's like a clock carried by everybody in the mind. I am not so sure it is the same thing as 'the biological clock', but certainly it is there, otherwise we could not know that one hour is longer than one minute.

However, now that we know there is no time permeating the universe of which we measure our version, how do we explain the phrase 'measuring time'? How do you measure something that does not exist? On closer examination, it turns out that we do not measure time; we create it in specific units, using the year as the base. The year is determinate, so it is one unit. Other fractions of the year, obviously, are also in units. And that is the end of the quandary. Even the atomic time is in units. We count the pulses or oscillations of atom one by one. Russell observed that time is now "Relation between points"---the year, for instance. I believe he was right. It is the logic of time in the universe. But it means the time is discrete, secular and known through how it is passing only.

The year, as our basic unit of time, is only a physical journey round the sun. However, through the sense of duration that we have acquired from the past and carry with us, it is easy to assign a period for the year (a long one), and specific sub-periods to its fractions. And that, I think, is how we can have time other than measuring it from the cosmos.[231] While the year on its own is not definable in logic, we have however used the day and night system and mathematics to give it length or duration such that so many days add up to one year exactly, and a new year begins.

So the truth is that the clock is no longer measuring time for us from the cosmos; we manufacture it to reproduce units of time obtained from the earth-year and deliberately programmed into it for reproduction. We are able to do this because the year has remained relatively the same under secular time.

This is the time that some people were determined to equate to space. No wonder that their efforts were futile. It is conceded that it can be done but not in actual physical reality, only in

---

[231] There is no method for stating the length of the year as a logical unit of time without using any of its fractions to define it---the nearest thing is distance or space.

mathematics alone and therefore of no use to logical thought.[232] For one thing, time in the new, secular sense is not 'time' in the old absolute sense, but how it is passing through nature only. It looks like the old time because the year has remained relatively the same---year after year after year, requiring only minor adjustments to our time in the clock based on the year. It is because a time system that is based on repetitive cycles is always the same, so long as it is related to the earth-year, or even the atomic oscillations, that Minkowski was able to get away (at least for a hundred years!) with his arbitrary formula for equating space to time. Einstein also could use the Minkowski formula because, again, the year and its fractions are relatively the same and he knew it. This has been a long digression. Now let us consider a few of the other aspects of time that always cause confusion.

5. Curved Space-time. Of course, curved space as the cause of gravity has been confirmed. It is the basis of many conjectures in cosmology. One of them is that as space and time are unified with the Minkowski mathematics, when space curls it will take time with it, making time travel a real possibility, so much so that one could meet his grandparents even before they are married! However if the Minkowski formula is fictitious and arbitrary, then this theory is also false. Thus, in my little book, The Coming Revolution in Physics, I suggested that because curved space-time is not a valid logical proposition, physics and cosmology, are heading for a major crash to shock the scientific world pretty soon by relying on it. At present no mathematician, physicist or cosmologist seems capable of thinking beyond the concept of curved space-time which they get from the Minkowski fiction of 4-D geometry. They say $s=ct$[233] and therefore space is the same thing as time and so since space is continually curving in the metric of general relativity,

---

[232] The pretence that Minkowski makes relativity easy to understand is like saying counterfeiters make the central bank richer. They do increase the number of bank notes in circulation, but is that desirable?

[233] This equation is cut short. I always cut it short since this is not a mathematical thesis or a book on physics. There is no point in reproducing it as originally presented when it is not going to help us in anyway. This is a book about time and how the 4-D geometry theory is distorting our most cogent view of how we get time. '$S=ct...$' means there is an equation that time is the same thing as space.

time is also curving with it. But the logical point they miss is this: in the strong gravitational field of general relativity there is nowhere anywhere for life to flourish. The grand parents would have nowhere to live. Yet Einstein divided the cosmos into two distinct categories in human perspectives. Of course he was not God, only another human being, but he was 'Einstein'.

It is sad that scientists do not realise this. Let me inform them that there is interstellar space, known commonly as the metric of general relativity, where gravity is so strong that there is nowhere anywhere for anybody to live and flourish and have opinions and children and also for civilizations to rise and fall. Then there are perspectives in or on inertial frames, fields, or planets, for short, where love and religion rule; children are born to lucky parents, marriages are contracted and annulled in daily cycles of love and hate, politicians and lawyers practice their evil art of influence, lies and deception and, finally, the taxpayer funds scientific researches to enable some mathematicians, physicists and cosmologists to live a good life and go on to theorise falsely that s=ct and therefore curved space-time makes time travel a 'scientific possibility' to delight the false prophets of religion.[234]

6. "Time Zero and the beginning of time." There can be no beginning of time if the time is discrete time; alternatively if time was not started by some supreme being from a specific date

---

[234] Talking about lawyers and their dark arts for defending the indefensible, I recall that I once went to the Royal Court of Justice in a litigation case, where I was suing without a lawyer. The lawyer for the opposing side and judge came from one chamber and were gossiping all through the hearing whenever the tapes were off. The judge even abused me to the lawyer, saying, in one of many abuses, "Of course you'll get the decision, but you know that cost against these creatures is academic." I had no doubt that he meant it as a racial insult, and said it right over my head to the lawyer behind (in litigations at the Royal Courts of Justice, Litigants in Person sit on the front benches.) He also let him draft the questions to be put to the jury by himself and typed it from his chambers and submitted it in court for the judge to say, 'that'll do'. He tried very hard to get me to commit contempt all through the hearing. He was later given a damning criticism by the Times for his conduct in another case, and later sacked, although they made it look like a resignation/retirement---appointed by Tony Blair in his forties? Unlikely. He's no good and they booted him out. That was one revenge for me; at least he and his racism are no longer on the Bench. This note is my second against that bad judge in as many books, not bad. I am satisfied---got my revenge!

and set it running all through the cosmos at the same time and the same everywhere.[235] The statement must be changed to "The Beginning of existence"; otherwise the relativity notion that every inertial frame has to have its own time is breached: time is based on matter, not matter on time; astronomers and cosmologists have themselves hopelessly confused about this one. They are following and reciting the myths of the religions instead of leading them out of the mythological world of their imaginations. Otherwise, what in Heaven's name is "Time Zero"?[236] And why are scientists discussing it so seriously as if the whole of cosmology depends on it? Time Zero from when---and how do you define that 'when'? You need time to determine that; and what system of time gave that period from which Time Zero began? Are we to take it that it began from "The Dawn of Time"? If so how can we fit relativistic time into that system of time---or what is the dawn of relativistic time, the birth of Albert Einstein, perhaps?

Time is not the same thing as existence. As explained in this book, you have to quantify existence to get time, or to get something that you can use as 'time', but that would merely show how time is passing and never what it really is. The result of how we obtain this time is that it is in determinate units only, the years and its sub-divisional fractions, for instance.

Similarly, there can be no end of time; but if the earth (or any 'body') ceases to exist then, of course, its time system will also disappear, because the time is based on existence, or matter, not the other way round. That was the religious view, but it is no longer tenable under the relativity 'fragmentation of nature'---see the kinematic sections of Einstein's June Paper.

It may be legitimately asked, "how is time based on existence?", and not the other way round, in the sense that time gives living organisms 'natural periods' during which to exist---on the grounds that the natural periods are determined by one's inherent physio/chemistry? The answer is that they are one and the same thing or complementary. Physio/chemistry will determine how long you (or anything) will live; and that translates into time units, say years.[237] The point

---

[235] This is what makes the Einstein notion of time novel.
[236] If time started from a certain date, how did that date come about before the time started?

is, there are no years in nature naturally without sentience; in other words, there are no time units. We created them by counting regular cycles as units of 'passing' time, like the years. Sentience creates the time units, or time, out of the physio/chemistry of being there. It is conceded, of course, that we create this time out of elements existing naturally---but in the absence of somebody to set the points and count the orbits of the sun as years, there will be no years and no seconds.

Once we rule out the influence of God and our own fallible and human, false and devious, religious gurus claiming to speak for God, we will discover that a system of time for any planet is bound to be an invention of human beings for cultural purposes, and that the natural time upon which it is based is neither definable nor even usable as time for the purposes of creating civilizations, science and technology.[238] I must stress that this notion of time is extremely important as it goes to the roots of all existence. No life anywhere in the universe can have a different time system---it is utterly impossible. The delusion is this: existence appears deceptively to be synonymous with time, because enduring implies 'passing time'; but it is not the sort of time we could regard as culturally usable. Even then, using time is not the same thing as time. The time is one thing, using it is quite another. Only quantified time is usable. Logically any life anywhere in the cosmos will face the same quandary: it will have to quantify time. This means it will have mathematics; so if we develop our mathematics strictly logically, there will be similarities between this world and other civilizations in the universe.

The whole of life is based on time. What happens to us after we are born; but mechanised time, which is what we know as 'time', arose out of human needs. We have learnt to quantify time. And as explained above, we quantify time by the use of regular or repetitive motions---the year as pared down to its familiar fractions, for instance. (We produce socially convenient

---

[237] This has been discussed in one of the essays in this book under the title, "Time and Quantified time". Being is time, but without quantifications by means of mathematics in conjunction with astronomy or features of the earth, being per se will not give you the years and seconds---sentience is required; the ability to count is necessary, and a theory of numbers completely indispensable.

[238] See how the year is defined under paragraph 4 (c) above.

quantities or units of time for the clock.) But quantified time merely gives us the physical rates of the passage of time by means of cycles---the years, for instance---and never the true metaphysical nature of time which we can never know beyond the physio/chemistry of existence, or why things last for some periods. To endure is to spend time. If time does not exist, why do things endure? On the other hand, it is utterly impossible to define this time beyond the physical constituents that enable them to last or endure. That is why the nature of time is unknown---but we can tell at what rate it is passing by through counting the years.

So there are only two things in existence. They are life (including all material reality), and the time for its regulation; for unless a person dies immediately after birth, his continuing existence makes demands on nature according to time. Because all notions of time are pragmatic---we know of it through the way and manner it affects us. Thus we have learnt from history and anthropology that for centuries people died immediately after birth, given the fragility of human infants. Ignoring physical protection for the moment, to survive, you need food. This means after birth food is required 'after some time', that is when you start to feel hungry. That brings time calculations into the human set-up, or mind, for the first time. As the infant eats, he grows---again, over time. It takes time for anything to happen. That is natural time, since it occurs without any human intervention or planning. As we eat and grow we age, also 'over time'. Soon the knowledge that men need food to survive after birth, and after that, to grow and age evolved. Following these guidelines people survived, grew and aged; as they did so social life eventually came to exist.

Once society evolved, there was the need to know when to feed infants, children and adults, to enable them to survive, grow and live their lives in society. This is how culture, or existence, comes to require time for the planning and regulation of human existence. Eventually the clock was invented to reproduce units of time obtained elsewhere—e.g. from the fragmentation of the earth-year, and so forth. Any regular motions or cycles can therefore be used for reckoning time; they will show how much time is passing by counting the cycles. For instance, we think ten years have passed (as units of time) when the earth orbits the sun ten times. But in reality the ten orbits only give us the rates of the passage of time because they are mere physical orbits! Hence, Bertrand Russell's query regarding what is measured by the clock if cosmic time is abandoned is not correctly phrased. The clock does not invent the units of time it produces; if it did, the units would change automatically to account for changes in the earth's orbits, as stated above. The clock reproduces units of time obtained as fractions of the earth-year and deliberately programmed into it. That is the reason the clock has to be adjusted when the orbits of the sun change. The clock itself cannot do it. And the clock manufacturer, too, has to get it from somewhere. The inventors of absolute time never could indicate the source of time; they couldn't but now we can do so without mentioning God.

7. I come back to the question of history again, for history is the source of most of the myths and legends of time, particularly when it is added to Past, Present and Future. Under discrete time, there is no such thing as 'The March of Time'. History is not the march of time but the march of events. Only events can have antecedents and consequences that go on and on---always with their associated 'dates' of occurrence. It is simplistic to count the years and say they show the march of time---that there were ten years under this King or Queen, President or Prime Minister, but now there are twenty years since he or she was driven out of power or something like that. There are only two things in history: events and their dates. Even school boys can identify which of these can have antecedents and consequences that can march on forever. The dates are merely added to the events as they occur.

8. Can gravity affect time? It all depends on how the time is defined. Because of religion time was taken for granted. Nobody attempted any credible logical definition. Even Einstein did not do so. He showed that it cannot be fixed; and if it is not fixed we have to investigate how we get our own time; and concluded that every inertial frame has to have its own time---yes. But how is it to be defined, or what is it? I am having great difficulties getting anybody to look at my work because everybody thinks he or she knows what time is. But, in reality, we've all got it wrong. Hence the many myths and legends. Now scientists are getting in the frame with questions about gravity and time---both must be defined first. If time is not mere

motion, then gravity is not relevant at all. Gravity affects motion, yet time is not mere motion. On the other hand, if it is not 'Being', just being in existence, and assume that time is going to make us age automatically (without knowledge of what I call 'Quantified Time'), gravity (as recently defined by Einstein), is not relevant either. However, let's look at some of the facts circulating about this matter. The scientists are so convinced of this (due to the Minkowski formula) that they regard any questions as evidence of ignorance.

The experimental results about gravity and time, assuming they are accurate, merely demonstrate something like "The Clock Paradox" which Einstein said cannot be used to contradict relativity because of the effects of acceleration on one of the clocks.[239] Even then it all depends on how time is defined: 'Being' is not time per se because time requires points without which knowing "when to when" is not possible; the general passage of existence is not time, because there is no such thing as "general passage" in nature; all motion occurs in individual perspectives; if all of us were blind, nobody would notice any motions except our own crawls. Additionally, motion is not time because it is multitudinous. In fact, even existence itself is not general; there are myriad of things and animals, each existing uniquely by its own chemistry; no two entities of whatever description are identical if there is space between them. The principal basis of the human mind are space and colour differentiation for individuation and coloured shapes; that is how words are written, and it is the reason we can read and write, or acquire images for ideas and thought. In the words of Bertrand Russell, "The data of sight, analysed as much as

---

[239] I just don't know how anybody can take time near a black hole; yet we are told that time is slowed down by black holes. If it is a clock, who carried it there without being sucked in and thankfully lost forever? In any case, it would come under The Clock Paradox and subject to the same objection as voiced by Einstein. (We have to accept that, like the sense of duration, many of the legends and habits formed under absolute or religious time are still with us, and may actually be useful, like the sense of duration.) If motion is mistakenly assumed to be time 'marching on', well, motion is not time because it is not uniform and also it is multitudinous. If it is 'Being', we still have to agree that 'Being' is not time because time requires points and therefore sentience. You have to carry it there in a clock, in which case The Clock Paradox applies.

possible, resolves itself into coloured shapes…and sight is the principal source of knowledge of the external world.[240]

Scientists cause confusion and distort physical theory when they refer to time as some sort of a general thing in existence (that in cosmology some processes speed time, slow time or reverse time), ignoring the fact that under relativity there is no longer a universal time running through all the cosmos to be so affected. Multitudinous motions will not give you "when to when" as time without 'use' of points---a human invention---to quantify the passage of time by means of repetitive cycles. Otherwise we cannot even define time.

In any case, gravity cannot affect discrete time, alternatively defined by Professor A.N. Whitehead in his book, The Principle of Relativity, as "A time system is a sequence of non-interacting moments"---year after year after year, pared down to its discrete fractions, for instance. If curved space does not take time with it when it curls, as proposed in the Minkowski 'arbitrary and fictitious' formula, then gravity, even in a strong gravitational field, cannot affect time, and especially discrete time.

In a world of discrete time, the units of time cannot persist because they are passing by---and that is precisely what earth time is in practice. We know our time from the way it is passing by only. Does the second exist, if so where is it? Obviously it does exist, because it does occur, but does not persist since it is merely passing by---second, second, second---the only way we get to know of it. So are the years. The year ends on 31st December. It does not persist; it is rather passing by; therefore it is succeeded by another year instantaneously. It is 'a moment of time' however long it may be. So the fact that one year is long is immaterial; the principle is the same. Take the year as some sort of a long snake whose length prevents it from passing in an instant; but if parts of it are passing then it is passing. That would be the view in law, too. Similarly, even the second can be subdivided down into millions, all passing. They do not stop. Nothing about time is static. For all we can ever know of it is how it is passing by. As mentioned above, Bertrand Russell is wrong to speak of what is measured in a clock; we do not measure time because we do not know what it

---

[240] Bertrand Russell, The Analysis of Matter, Ch. 4.

is. The notion of time's measurement has been the source of many mysteries—but it is wrong. All we can do is to indicate the passage of time in specific cyclical units---the years, for instance---by means of repetitive motions. The atomic time mechanism is based on the same principle, and, in any case, the atomic oscillations have always to be related to the seconds to make any meaningful temporal sense. My conclusion is that gravity cannot affect time except in a strong gravitational field. Yet in such a field what we can use to track time would not be there; so there could be no time in the ordinary sense like reading it from the clock---unless earth time is applied in contravention of the Einstein theory of frames.

9. Yet, in another sense, Russell's query must be answered (that is the nature of logical reasoning), namely, in the absence of a universal time, how do we get time to programme into a clock? If all we have is the year and its fractions as "passing time", then all we can ever know of time is how it is passing by in practice, and that only gives us discrete time with all the peculiar consequences mentioned above---and more.

To recap: the true nature of time is unknown and it would appear that man can never know it. That is no reason for supposing that time does not exist; it makes publication of Railway Time Tables appear to be an exercise in stupidity, or even fantasy. It might as well be asked if the trains of the Time Tables exist at all. However, what we use for time is the earth-year. But the earth-year is just a physical journey round the sun. We count the orbits as 'years' to give us a measure of the rate of the passage of time and mentally assign periods for the year and its many specific fractions, including the oscillations of atomic time. This process cannot tell us the true nature of time. It is however sufficient to guide human life and culture. As such, evidently, time requires sentience, and the ability to count is also essential. Somebody must be there to count the orbits of the sun as 'years' or there will be no years and no seconds derived as fractions of the year. When the year is subdivided down to the seconds and other fractions we are able to know how safely to regulate life in accordance with the essential features of the globe---so that we avoid going to the farm or bush after midnight but rather during daytime; even then it would be regarded as safe only between certain hours. The units of time are important; they keep us safe. Without them the mere mention of time would not carry the serious meaning we get.

The most important aspect of this time is the fact that, since it consists of units, it is essentially discrete. We see time going year after year after year, or in seconds, hours and so forth.[241] However, discrete time carries momentous implications, most of which undermine our most cherished legends of time, legends that owed their power to man's primitive ancestry and instincts. Some thinkers are inclined to add scientific mysticism to our basic instincts about time with theories of time travel, and curved space-time. All this, after Lorentz and Einstein have helped man to free himself from the oppressive tyranny of absolute time. Luckily, it is now logically possible to analyse time consistently to the conclusion that it is neither general nor absolute so that a second here is not a second everywhere else; and the methods we use for this have shown that time is discrete and therefore the most serious and ancient problem of time, being its passage, can now be conclusively resolved. The passage of the years is the passage of time, and since there is only one year replicated to be years, the passage of time occurs through the procession of time units, which are, essentially, just repetitive physical motions we count as the rates of passing time. A sad admission that man can never tell the true nature of time.

In my opinion Godel was only partially correct. He could have proved without any doubt that time does not exist at all. In fact, it does not exist anywhere in nature;[242] yet there is something called 'time' which we have to use in all activities so much that it virtually dominates human existence; but on logical analysis it turns out to be just the passage of time we can ever know, and never the true nature of it. There is a world of difference between the two ideas, as I have argued all through this book.

---

[241] This is very easy to understand, for we call the passing years 'passing time', and count our ages as the number of years past. Yet the year is just a physical orbit of the sun; it can only show the rate of the passage of real time the nature of which is unknown other than the suggestion that it is probably chemical processing that forces on us the periods of waiting we call 'time' for want of a better word.

[242] What cannot stand still is not existing as such but only passing by. To exist and to merely pass by are two different things in metaphysics and epistemology as well.

So, in the end, we have to accept that there is something called 'time' but it turns out to be only how it is passing through nature, and never what it is in physical reality —which is something nobody can ever know.

The belief that gravity affects time is linked to the Minkowski formula for 4-D Geometry. If time is the same as space then when space curves it will take time with it in a gravitational field. But the theory is confused, quite apart from the fact that the Minkowski formula is logically untenable. First, there is no time in a gravitational field at all. Time requires sentience to construct, as Russell put it, and I doubt that anybody can live in a gravitational field to construct his time. The valid logical theory is what has been quoted, stated by Professor Jeremy Bernstein in his book, Albert Einstein and The Frontiers of Physics (Oxford, 1996, p. 110.) "In the absence of gravity, space and time are distinct entities. In the metric of special relativity they play distinctive roles. But in the presence of gravity the metric is altered, and space and time become mixed up with one another. The metric has four coordinates, but the space and time coordinates become entangled..." This is the theory but it is wrong. If space is not identical with time then it cannot happen.

The irony is that even if space is identical with time, it still cannot happen: both conclusions are wrong. How can space and time be one thing as proposed in the theory of 4-D geometry and yet get mixed up with each other again? It means one entity is conflicting with itself, which is illogical. So, in either sense, how can space affect time so much so that as the space bends in a gravitational field time too will bend with it? Yet, as the first part of Professor Bernstein's statement shows, in the absence of that space is space and time is time. The whole theory is misconceived.

Despite all this, some researchers try to measure the effects of gravity on time. The common mistake they all make is that they regard the second as existing independently in the universe and base atomic oscillations on it to measure changes in the duration of the second under gravity then call the results as 'changes in time caused by gravity'. They get away with it because philosophers fail to scrutinise their methods, since our modern-day philosophers are also afraid of relativity! In Britain it is a common pastime to declare ignorance of science and mathematics. Let me remind them that there can be no time in a gravitational field at all. The clocks they carry there is showing earth-time; yet this earth time cannot be applied anywhere other than the earth. Furthermore, their clocks must be suffering from the effects of the condition known as "Clock Paradox", which, as Einstein has explained, cannot be used to undermine relativity, because one clock is affected by gravity (or acceleration as he put it), but the other is not.

Finally, it should be clearly understood that there is no universal time. So our time has got to be deciphered; and we have come to the conclusion that it is 'constructed' by ourselves by means of repetitive cycles---the orbits of the sun as years, for instance. Secondly, in special relativity space is space and time is time. We still live in a special relativity frame. The conditions of our lives have not changed since Einstein. Thirdly, the Minkowski 4-D Geometry is based on imaginary time coordinates for the creation of time as 'space-time', which is illogical. It uses time to create another kind of time. Where does the first time come from? So his formula is regarded as arbitrary and fictitious. Fourthly, and more seriously, the second is not existing in nature as an independent time unit; it is rather a fraction of the earth-year and cannot exist without the year, yet it is experiments with the second that create all the confusion about time and gravity.

Being a fraction of the year, only the altered motions of the year (that is, through the altered orbits of the sun) can cause a change in the second's period or duration. Whatever mathematics is used, gravity cannot effect any such changes because it has nothing to do with the creation of the second per se. Since there are no universal, absolute or standard units of time, there can be no comparisons of seconds to show that one is either longer or shorter than the other, or any other second. To counter this it is claimed that atomic clocks are used to show that gravity can affect time. This event, if true, can only be discussed under the heading of 'Clock Paradox'. No clock can be used to show alteration of time, because no clock controls the whole of time, and there are no standard units of time by which to judge the performance of any clock. Any clock affected by gravity is performing badly, which is a matter for consideration under the Clock Paradox syndrome; and Einstein has decided that it cannot be used to undermine relativity.

Altogether, the subject is not even important in physics; it is used merely to support the Minkowski four-dimensional space; yet his theory is not tenable because it is based on imaginary quantities. The whole debate is sheer humbug, normally pursued by religious scientists, and they deserve to be ignored. However, a detailed refutation such as I have attempted is always required to defeat the detractors of science and their scientific mysticism.

REFUTING MINKOWSKI PROPER

There are several logical points for refuting the Minkowski arbitrary theory for equating space to time. A few of them are given below. It is true that the formula as originally presented is purely mathematical---and even Einstein praised his mathematical acumen. But the refutation here is not mathematical because the basic logical premise is written like this $\sqrt{-1}.ct$. This little group of mathematical symbols has ruled theoretical physics for the last hundred years. Yet, translated into linguistics, it means time is replaced with an imaginary time, due to the i in the equation. It is the logical basis of all his mathematics and so it is the one I am interested in. Hence my objection is linguistic not mathematical---in plain language, there is no such thing as imaginary time, so to base such a powerful equation on imaginary time is wrong. Even Professor Sir Arthur Eddington called it arbitrary and fictitious. I'll settle for that. Mathematicians are fond of inventing notions, quantities qualities and physical properties then impose them on all of us on the planet through their theories. We cannot accept that practice for the determination of physical reality as to whether or not it is four-dimensional. In the absence of a better proof, we must conclude that space and time are separate entities and therefore every supposition based on the Minkowski *formula is flawed. The wonder indeed is that since Einstein, Russell, Whitehead and Eddington, intellectually we seem to have entered a new world of cowardly thinkers who are literally afraid to stick their necks out. Eddington said the Minkowski formula is fictitious. I often wonder why nobody since then has seen fit to remark on it but just use it as if it is logically valid? There is no shortage of brain power. Rather I think scholars are too busy making money from the media.*[243]

This is it then. Perhaps refuting Minkowski deserves a full chapter to itself, although personally I don't think so because all through the book I have tried to show that his formula for equating space to time is logically untenable.[244] But just in case the reader would prefer to have my grounds for rejecting his theory grouped together, I give below the main reasons for rejecting a theory that has been declared even by one of the leading experts of relativity, Professor Eddington, as fictitious.

The Minkowski theory is not a major philosophy to tackle it so seriously; strictly speaking Minkowski is not even relevant in the serious discussion of time, except that in science his terms (particularly 'space-time') are continually employed with connotations that are logically unacceptable; and also his formula deals with time, the second most important subject in human life. And the way it deals with time is also philosophically unacceptable; it must be refuted root and branch.

My basic argument is that it is not correct to state that time does not exist on earth; at least we know how it is passing by because we use repetitive cycles to keep track of that, and the repetitive cycles or motions make it necessarily discrete so that, as a bonus, there is no need to find any theory about how

---

[243] Undoubtedly this is the age of television. Everything is organised for the benefit of television producers and viewers and it seems we will die with it---science, the search for truth in philosophical scholarship, international politics and diplomacy, including all economic issues and all wars, human welfare and the purpose of life; all these are now either good for television or not worth pursuing.

[244] But there you are. In writing a book of this nature one has to think of three interested parties: the reader, the critic and the scholar who is mostly interested in what has been written down.

it is passing by, since all we can ever know of time is its passage through nature and never what it really is. And Minkowski has to be shown to be wrong because if he is right then my argument rather is wrong and time is everywhere---that is, in every space. Yet my argument cannot be wrong because we actually know only of time's passage---the years, for instance. Or even the atomic oscillations. We know them after they occur, not before they occur; but their occurrence is the passage of time. So long as the occurrence is successive, it means the time or what we regard as time is passing.

I must stress again that Minkowski is not intellectually important; every reference book describes space-time as artificial. You cannot go to the moon by means of an artificial theory. But he has to be formally refuted because his theory is distorting relativity and physics as a whole since all scientists refer to every space as space-time. Whatever they say to the contrary, the truth is that scientists seem to believe that space and time are linked because they have failed to define time. They are linked, of course, but only in the sense that we cannot have time without space but not in the sense that they constitute one entity. For we have to use points in application to space to get the earth-year as our basic unit of time out of which all other units are derived. That alone shows that the time we obtain is independent of both space and the points. We all know that there is no longer a universal time. So we have to trace the origins of our time (whether human or divine), but scientists seem to have settled on the Minkowski space-time concept absolutely. If it were just an alternative notion of time we could learn to live with it as one of the many interpretations of time. On the other hand, since it is proposing that space is four dimensional, it changes the complexion of all theories and has to be shown to be false---because that is what it is.

First, let me point out that Minkowski was not trying to solve any problem in physics and therefore deserves no credit for his efforts.[245] Normally, every scholar or scientist is given his due

---

[245] I know, of course, that Minkowski can never be refuted with arguments, not even with mathematics. He is as strong as religion for three reasons that have nothing to do with truth but more to do with the crazy manner the human mind works. (1) His theory is closely associated with God (Einstein). (2) Everybody calls it 'artificial' and still believes in it. (3) Mathematicians think it makes it easy to understand relativity, and although they are wrong,

for his efforts---but they must be recognised as desirable efforts. Take the eather debacle for example. Thousands of scholarly papers were published to try and solve the problem of the propagation of light. All of them failed miserably; yet we recognise that the writers were commendably trying to help us solve an existing problem; they were therefore worthy of our gratitude.

But in the case of the Minkowski proposal to equate space to time, there was absolutely no problem obstructing physics in a similar way. He rather arrogantly and insultingly called Einstein 'a lazy dog' who did not take his mathematics lessons seriously when he taught him at the Polytechnic; and said if he had done so he could have seen that there was a mathematical method for incorporating time into space naturally to replace the 3+1 formula since it was not completely objective. And, of course, it is true that if time is to be recognised as one of the coordinates for defining physical reality then time must be objectively established not added as in the 3+1 formula which must depend on human intervention.

This much is understood. But it was not a problem in physics. The philosophers may worry their lofty brains about it, but it was not obstructing anything in physics as the propagation of light was in the eather debate. Minkowski took it upon himself to produce the mathematics necessary for equating space to time, and told the world that from the moment of his lecture inaugurating the theory, time and space ceased to be separate entities and became one. What vanity, since mathematics creates nothing but only reveals what is there. Yet as it turned out, the 4-D geometry was not there. His mathematics may be beautiful but it is not accurate because it is based on flawed logic—i.e. imaginary time coordinates. However much the transformation of coordinates is defined (or manipulated, extended, juggled and dressed-up in bemusing counterintuitive mathematics), time cannot be physically incorporated into space unless by magical 'jumps', saying something like, 'we

---

they snoop all over science and are so vindictive and ruthless that everybody is afraid of them--- with the exception of myself, Einstein, Bertrand Russell, and Professor Sir Arthur Eddington, all of whom are now sadly no longer with us, and I am also a very weak man in the mid-seventies. Yet of these mathematicians even Newton was afraid of them. The subject is so intricate and wide ranging that it can be used to undermine any theory.

can therefore assume that time will become part of space', which is not a logical way of making it so.

What Einstein was proving is that time can only be obtained from space, and that the Lorentz local time concept may be taken as 'time, pure and simply', meaning that it is the only way to get time everywhere---by your own 'local' space, so that every inertial body can use its local space to create its own time. But you cannot use the transformation of coordinates to incorporate the time into space naturally because we even do not know what it really is, as we can only know how it is passing by.

So the Minkowski theory remains a mathematical proposal without support (or substantiation) in physical reality, defined as what can be recognised to be in physical existence not what the mathematician says he knows to be there through his mathematical symbols alone. In the matter of time, we want it in a form recognised by even uneducated, primitive people as 'time'. The academics may not agree with me, but I still think time alone is not an ivory-tower subject. What has happened over the centuries of thinking about it is that religious people, mathematicians and mischief-makers, being ignorant of what it is, have nevertheless managed to wrap it in stultifying mysteries than it deserves. It is strange, of course; but let's face it, Lorentz and Einstein stumbled across its secular nature either by luck or genius, so now (at least) we know it does not come to us from the universe but that we create it.

When they go to the farm people want to know when it is time to head back home safely without mathematics. Above all, time was never properly defined by Minkowski. His attempted definition by use of the term 'interval' was condemned by Russell. When time is analysed logically down to the bottom, it becomes discrete not something permeating all space to be invoked with the appropriate mathematics. I list below my point by point refutation of the Minkowski theory.

(1) Time is central to the Minkowski formula, yet it is only invoked with an imaginary time co-ordinate. This is logically untenable because there is nothing imaginary about time. It is so real and oppressive in all human affairs that we dread its passing as it brings ageing, associated with infirmity and death, or, at the social level, missed opportunities.

(2) It is evident that Minkowski failed to define time in clear terms as to whether it is universal or discrete.[246] We are certain that universal time does not exist (time dilation, the concept of local time as time, Einstein's analysis of order and simultaneity to show that no unit of time is universally applicable, together with his theory of frames all confirm that "there is no longer a universal time".) Yet the Minkowski theory seems to imply something like a universal time that can be invoked with the appropriate mathematical symbols.

(3) Minkowski said plainly that time and space are unified from the time he announced his theory. He actually stated that 'Henceforth space and time constitute one entity, etc.' It is true that under relativity space and time are not seen as fixed but rather as dynamic, however the concept of local time which Einstein transformed into space-time to the effect that every inertial frame (local space) will have its own time implies that time is obtained from space with the application of points to space, in the manner we get the year and all of its fractions. This means time is a product of space and points; therefore it is contradictory and against logical reasoning to claim to be capable of linking time to space to form one entity. As the product of space, it can only logically be added back to space, as we know in the 3+1 formula. A thing cannot be produced by space and yet be naturally and inseparably the same as space again. They are related, certainly, but not one and the same thing.

(4) The Minkowski ict equation mentions time twice---as I have said before. The i is supposed to invoke imaginary time, yet that is multiplied by ct, being the velocity of light and time to create space-time as a continuum incorporating time in space

---

[246] *Let me stress that time is discrete precisely because of how it is created; and being discrete means it passes by through the procession of its units. The basic thesis of this book is that we count the orbits of the sun (as years) or oscillations of atoms for time. But they are passing, so we only know the rates of the passage of time not what it is (it may have something to do with chemistry, but that is another matter.) We cannot count the oscillations of atoms before they occur. We count them as they occur---that is passing time. In the same way we count the years as passing time for ageing, for instance. You do not age ten years before they occur. You count the passing years to get the ten years. But that is how time passes and therefore no theories of explanation for the passage of time are required.*

naturally. I am at a loss to know which other time again? Where does it come from? On this issue, I am inclined to abide by what Einstein himself said. He is the master of the whole idea; and, as quoted from his book, he states clearly that i is supposed to represent the usual t for time. So how does Minkowski multiplies the i by ct again in the equation for the purpose of creating his space-time? Where is the extra time of the ct coming from?

(5) As Professor A.N. Whitehead has pointed out, time and space still pass through nature separately, and he was an outstanding mathematician, so nobody can accuse him of failing to understand the Minkowski mathematics.

(6) If space and time are naturally linked, we could never have had time on its own, or space on its own as the situation was before Minkowski---and still is, as Professor Whitehead has pointed out.

(7) Over the 50 or so years that I have been working on the problem of time, and sending many technical papers to the academic journals, I have encountered many insults and references to elementary mathematics almost all of them in defence of Minkowski, so much so that I once wrote back to advise one referee that at this level his contribution was so elementary as to be an insult to my intelligence! The ds^2 formula, the square root of minus one, the Lorentz Transformation formula, and his equation for his Esynched clocks $t^1 = t -- Vx/c^2$ have all been mentioned to me, sometimes by those referees of academic journals who know nothing beyond formulaic concepts and ancient mathematics. They often fail to remember what happened over the eather debate before Einstein. And they believed that mentioning these old ideas would put me off. However, being simple in nature, if not in mind, I always go back to the master Albert Einstein himself. He gave us no complicated mathematics. He just said, and I quote again, ".... we must replace the usual time co-ordinate t by an imaginary magnitude $\sqrt{-1}.ct$..." That is all a logician or philosopher wants to hear. For the point of contention is this: the basic Minkowski mathematics equating space to time is called 'ict'. It relies on i---as an imaginary time co-ordinate, exactly as Einstein put it. So according to Einstein himself it is an imaginary time co-ordinate. But there is no such thing as "an imaginary time" to have its own separate co-

ordinate in the determination of the nature of physical reality. This formula is obviously wrong---no one should try to defend it.

(8) What really amazes me is that mathematicians overlook the fact that linking space to time is intellectually more important than the discovery of the cause of gravity by Einstein; and yet, while they are prepared to go to the African jungle to test the Einstein theory of gravity, they accept the Minkowski formula on the logical merits of 'thinkability' as even Einstein himself put it (perhaps jovially since we now know that he did not believe in it), instead of subjecting it to the most rigorous logical scrutiny. Yet it takes about just a moment's reflection in logic to realise that imaginary time co-ordinates cannot support such a major shift in human conceptions of time, space and the nature of physical reality. Space-time as 4-D geometry, if true, must be the greatest scientific theory of all. Yet even Eddington told us that it is just not true of the physical reality we have to deal with in physics and technical philosophy. Even by saying its proof is something to be regarded as thinkable, Einstein inadvertently exposed the theory as logically flawed, for how can people's thinkability be part of the factors that determine physical reality? Good lord, we are described as 'two-legged reptiles crafty and venomous' for very good reasons---so bad that if 'thinkability' of people were to be relied upon in physics, even the cemeteries won't accept us for burial![247]

Furthermore, the way Eddington put his mild condemnation of Minkowski was not good enough. As a book entirely devoted to the mathematics of relativity, Eddington had a duty in his Mathematical Theory of Relativity to inform readers as to why the Minkowski mathematical interpretation of relativity was arbitrary or fictitious; for these are, after all, quite serious condemnations; and this is a very serious matter, too, for relativity is a completely novel theory of physical reality, and it has been proved experimentally. To try to interpret it mathematically only to end up promoting fictitious theories is a very serious charge indeed.[248]

---

[247] It is even possible that he was mocking the mathematicians with it, who knows?

[248] It is not enough for Professor Eddington to say the 4-D geometry is organised (or drawn) like the latitude and longitude lines are drawn on earth. We are living with these imaginary world lines without encountering any problems in physics, but we cannot live with the 4-D geometry without

Parts of special relativity are philosophical because of time. Thus some of his peers called Einstein 'philosopher/scientist', and all the evidence is that they were right. As a result, the whole of special relativity cannot be rendered in mathematics. Lorentz saw this, as he later admitted that he failed to discover the special theory of relativity because he did not take what he had discovered about time seriously. That discovery was 'time dilation': to the effect that a speeding clock appears to run slowly. Deductively it leads direct to the concept of local time. Your local time runs for you normally; but it will be seen as running erratically by people looking in from outside your locality.

My own opinion is that, apart from the two postulates and the Einstein theory of frames, Lorentz could, indeed, have discovered special relativity, but because of them he never could have done so merely by taking his own recently discovery seriously. It is partly that discovery Minkowski referred to when he said his theory of time had 'sprung from the soil of experimental physics'. The other part was Einstein's analysis of order and simultaneity. Since he was a very good mathematician, Minkowski must have noticed the difficulty with time in making special relativity purely mathematical, and, I think, decided to get round it with the suggestion that time and space constitute one entity. In other professions this might be called fraudulent; here we can only observe that it is illogical simply because one cannot cheat nature.

9. Minkowski never defined time sufficiently to make him worry about its passage through nature. The nearest he got to was the term 'interval', which is condemned by Russell (see chapter XXXVIII of his Analysis of Matter.) In all science time is never clearly defined. It is spoken of as if it is so familiar that it requires no definition; yet it ought to be clearly defined before the mode of its passage can, properly, be determined. Lorentz and Einstein have shown that our time is our own local time. How did it begin? What is its essential nature? How does it pass through nature so that the years grow to become the

---

causing problems in physics and philosophy. In physics because it has to be true else it contradicts special relativity; and also in philosophy because it seems to make time universal again ---something that can be invoke with mathematics, rather than something that can only be created locally for local purposes.

centuries? Attempts have been made to logically answer all these questions in this book to the best of my ability. That ability may not amount to much; but that is the more reason to make what I am saying clear to everybody. For the ferocious, cruel, religious critics are ever eagerly watching we heretics. I believe they can't find much wrong with what I am saying. For instance (let me give one crucial example), as quoted before, Einstein in his book "RELATIVITY" stated happily (happily because he was praising Minkowski in the book!) that "we must [(i.e. *must*] replace the usual time co-ordinate by an imaginary magnitude $\sqrt{-1}.ct$ proportional to it". This is astonishing to us now; at least it is surprising to me. We think Einstein was so clever that *if he had defined time logically in the absence of his own theory that cosmic time does not exist,* he might not, even then (in those heady days when Minkowski was King), have accepted the imaginary time coordinate so gleefully.[249]

Time has never been scientifically defined by anybody; the truth is that we can't, but people were not honest enough to admit it. It is (or seems) so obviously there that we all take it for granted; besides the mechanics have created the clock, and people just use it. But once any of its aspects or features is questioned---that it is not general or absolute as happened in the past---we ought to sit back and examine it logically, properly or critically. That is what I am trying to do in this book. I have to admit that the journals ignore me because they are totally committed to Minkowski and the term 'space-time' seems so sweet that nobody can even imagine that it might not be true. For, obviously, if it is not correct then I hope everybody will agree that physics is in big, big trouble, therefore the journals have to be cautious. In any case, normally the journals frown upon negative ideas, and they know that even Einstein not only accepted the Minkowski formula but actually used it in the field equations of general relativity. That is the reason physics seems to neutral observers like myself as hopelessly distorted.

---

[249] I plead with the reader to forgive this frequent repetitions; for this is the ultimate of theories, second only to the explanation of life and this is the best I can do. It's not easy. Yet I think I have an obligation to stress the essential aspects to make what I am saying absolutely clear so that the reader can judge it well.

Minkowski got away with it because relativity is not affected on this planet however we define time.[250] Professor Eddington could insist that "we shall continue to employ it." He is saying scientists or rather mathematicians shall continue to employ a theory he had accurately described as fictitious and arbitrary. The technical reason is that the Einstein condition that time should be a separate co-ordinate in the description of phenomena is fulfilled no matter how the time is defined. All time looks like universal time due to the common earth-year periodicities we carry in the mind. The difference is in how it is obtained: either it is created by man as 'relation between points' or it is assumed to be 'given,' a providential bounty, as they do in theology, implying the total rejection of the Einstein notion of time.

Another critical point is when the 4-D geometry (carrying time with it) is applied to other frames, including the metric of general relativity, since the periodicities are unique to the earth's cycles. I think mathematicians simply employ earth time from their minds in breach of the theory of frames. Let me explain what I mean: time has to be quantified; the earth's periodicities derived from the peculiar conditions and cycles of the earth's parameters are our sole means for doing so on this planet; otherwise the word time has no meaning. Time is always 'how much time', and we can only know that through the quantification of time.

Moreover, in practice, as opposed to theory, it is in the determination of how much time that problems arise. Hence the term 'proportional' is employed in the Minkowski equation. But proportional to what "standard unit of time"? Or proportional to

---

[250] As I have said before, I am certain good-old-clever Einstein knew exactly what he was doing with the Minkowski fiction; and he knew it was fictitious because Eddington had said so. Furthermore, he had denounced the mathematical interpretation of his theory, saying that because of the mathematicians he could not understand his own theory any more. But he also knew that Minkowski was hailed by scientists as the great mathematical genius whose formula had helped them to understand relativity. Besides, all they wanted was the rejection of the 3+1 formula. And he knew that whether time is 'space-time' and part of 4-D geometry or not, on this earth, time is always the same and the rejection of the 3+1 system could not vitiate his theory of gravity---which was the important thing as far as he was concerned.

what? To decide that we have to use the earth's cycles; the units of time used proportionally are limited to the earth and its cycles alone. So Einstein was right, for that makes earth time applicable solely to the earth. Otherwise we are transported back to cosmic time where units of time are there free for anybody anywhere to invoke, and the same everywhere. Thus the mathematics of Minkowski is not even relevant, since there is always time in every action, every event, and every situation. It is carried in the mind. What Einstein meant was that, having disproved universal time, and made time local in origin in a fragmented universe where time must be limited to a frame, he naturally realised that time must be physically added to phenomena to make the natural laws of a frame uniquely limited to that frame.[251] That is why his equations for motion in special relativity consist of the three natural dimensions of space, plus time---that is, plus our peculiar time units derived from our own space. If time 'is' universal he would have had no need to do so.

But Minkowski cleverly saw that time is always in the mind and so any symbol we use will be able to give us time in the earth-time periodicities to which we are accustomed. However he forgot that the time units to which his 'proportional time' has to be proportional to (the hours, minutes, seconds and so forth) belong solely to the earth; they are the periodicities derived from the earth's cycles (or parameters), and under the theory of frames, cannot be applied to other parts of the universe. How therefore can the 4-D geometry whose time (or implied time) is based on the earth's own periodicities be universally applicable? In other words, how can the 4-D geometry be relevant to the universe? The time that it will invoke is the very same earth time we all carry with us in the mind, such that any symbol we care to use can invoke it. Minkowski was cheating and he probably knew it---some mathematicians must have known that, too, since they refer to his theory as "The Minkowski universe", knowing it is false!

The rule is that in every frame the time appropriate must be one derived from indigenous cycles, precisely the way we get our time from the earth's cycles. For this reason a second here

---

[251] It is often forgotten that the two postulates imply that our physics is limited to the earth.

cannot be the same as a second somewhere else since the parameters would be different. Thus earth time should not be applied to other frames (including general relativity, I must stress again), if time is not a universal phenomenon. And we all know that the reason time is not universal is cast in solid logical concrete because of the insurmountable evidence presented by the event described as time dilation and the local time concept based on it, even if the analysis of simultaneity is disregarded, but I doubt that we are at liberty to do so. In my opinion, Time Dilation is one of those things that happen to mankind by accident to make life bearable. By logical definition, it simply is not true. Time can only dilate through changes in the earth's motions. So there was no time dilation at all in the Lorentz experiment; yet it led to the abolition of cosmic time with the implications that time is not God-given, not permeating all through the cosmos and the same everywhere, and probably does not exist, since what we call time is merely how it is passing by only.

Again, Professor Arthur Eddington said, "... we must endeavour not to lose sight of its fictitious and arbitrary nature", while continuing to employ a theory that he has described as arbitrary and fictitious. Where in the gamut of intellectual activities do scholars do that---that is, use a theory that is not true with whatever caveat? And why should it work? I believe it works because the time periodicities it invokes are already carried in the mind and are derived from the earth's cycles. But that is in violation of the Einstein theory of frames.

The question is why Professor Eddington has to say that the Minkowski theory is arbitrary and therefore could only be used with that caveat in mind---to what purpose?[252] If it is not true, it must not be used. But if he was going to continue to use it, what is the purpose of the warning about its fictitious nature? One reason, I suspect, is that people instinctively use the earth-time periodicities, since relativity is not affected however we define time. A second reason, I think, is that in another part of

---

[252] Professor Eddington did not even mention Minkowski by name---was he also afraid of the mathematicians? Quite frankly these were difficult times. Einstein was alive and he had praised Minkowski---yet it was clear that the Minkowski formula was extremely serious but based on imaginary quantities, and Bertrand Russell too was there, Professor Whitehead was also watching!

the same book (his Mathematical Theory of Relativity, Ch.1.8.), he observed that "The rough measures of duration made by the internal time-sense are of little use for scientific purposes, and physics is accustomed to base time-reckoning on more precise external mechanisms." So he was aware that time is carried in the mind with us wherever we go, and whatever we do.[253] And that is correct. In other words, the erudite Professor realised that time is not entirely mathematical, but partly philosophical, since it involves the internal time sense as the sense of duration, based on the memory's facility for retention, the sense of things lingering, or enduring. Because of this, time can clearly be seen as the union between the sense of duration and external cycles. Therefore it can never be rendered entirely mathematical. So he called the theory he was using 'fictitious and arbitrary' although it worked for him and recommended that it should be used with a weak caveat. But, philosophically, that is an admission that the whole Minkowski theory is illogical and untenable.

There has always been two aspects of time: the philosophical or psychological and the material or physical cycles we have to use to quantify time. The physical is subject to mathematical manipulation. The philosophical aspect of time is not. As a result, this Minkowski business is not a simple one, or an easy one. It is not just a matter of deciding whether to accept or reject his theory; it goes deeper than that.

We start with the special theory of relativity, where the equation for motion consists of four parameters: the three natural facets of phenomena, plus time. And since we cannot contradict the special theory of relativity, the debate is as serious as the one about the truth of the Platonic theory of Forms/Ideas. It is a vital philosophical quandary with implications for theology; but it cannot stand as part of physical theory because it rests on imaginary quantities. Even theologians and philosophers of the

---

[253] The problem is whose time? If it is earth time it cannot be used outside the earth---but they do, because relativity is not affected whether the time is considered general or local. However, in general relativity cosmologists are experiencing problems, so much that their suppositions seem bizarre. Furthermore, it will not be possible to link the quantum to general relativity while the time used in general relativity is not relevant to that frame, for the quantum is time-dependent. The time by which it materialises is earth time, relevant to the earth and its cycles alone.

various schools are exhorted by scientists to accept only objective reality as experienced through the senses without complex theoretical explanations of what is experienced. But, if so, then the first casualty is Minkowski, also a mathematician. His formula, so adored by mathematicians (the very people who affect to report on reality strictly objectively), cannot be accepted as part of the corpus of physical theory.

I want to end this rather boring section of the book with my own thoughts about the conflicting theories of Minkowski and Einstein concerning the nature of physical reality, as regards whether man contributes to the nature of physical reality with his time, as Einstein proposed, or time is intrinsically inherent in space already as Minkowski suggested.

I can see the merit of the Minkowski formula and appreciate why mathematicians regard it as useful---no more than that. Since Einstein had shown that cosmic time did not exist, the 3+1 formula suddenly acquired the aura of the most serious philosophy of physical reality for two reasons. (1) How do you get the time in the absence of a universal time before you add it to space? (2) Who is going to add the time to the normal dimensions of phenomena as a separate (or additional) co-ordinate? A role was reserved for man the observer to become part of the observed, which introduced a whole new facet of physical reality.

Minkowski disagreed; and he was clever enough in mathematics to be able to disguise his disapproval with intricate mathematics. He sought to answer both questions with one complex mathematical formula. The solutions, he said, come from making (or recognising that) time is naturally part of the physical dimensions of space. If relativity had been well understood at the time, logicians would have realise that the two systems of Einstein and Minkowski threw the philosophy of physical reality into a confused and dreadful intellectual quagmire; either way the world of sense was changed out of recognition.

I say this because it has to be realised that this is a discussion of the ultimate of physical reality. A subject that is not legally restricted, yet only a handful of humankind can understand it, let alone make contributions to it. To say it is difficult is an understatement. The reason that, sadly, up until now relativity is not properly understood, some say, by anybody. As such, in this rarefied area of human thought, we are only permitted to call something useful; we have no cosmic authority to declare anything as 'the truth' in physical reality. That is the role of logic, and logic alone.

Concerning the debate itself, Professor Eddington puts it best. He says of the Minkowski system that "...it is of great utility and convenience in describing phenomena..." That is true; but then it happens to introduce another problem (its own problem), so that we have two worries. One is the controversial Einstein 3+1 system incorporating a new conception (or philosophy) of human nature, and therefore only partly scientific; although everybody recognises that it is a profound theory with many merits never seen in physics before. The second worry is with the imaginary quantities of the Minkowski theory.

Einstein made man (the observer) part of the observed--- regardless of the metaphysical status of man in the cosmos which he did not bother to mention and rightly so---as he has to add the time to the other dimensions of space. One can imagine that from the Einstein point of view, no matter how man got here, and for what purpose, he contributes something to reality as we know it.[254]

Let me spell this out clearly: The three standard dimensions of phenomena are natural; the addition of time to phenomena as a separate co-ordinate is a conceptual novelty introduced by Einstein, but it has been shown to be correct. The only drawback is that it is additional to the other dimensions of reality. Knowing time as it is (or was known to be), namely independent of space, how was the addition of time to space as a separate co-ordinate to be achieved?

The answer is that a role was carved out for man to add the time; the observer became part of the observed. That is clearly not scientific, it is conceded. To be more scientific time must be shown to be physically part of geometry, if possible, as part of

---

[254] Of course man is part of the observed; for reality is what we see; but what we see is only what our defective senses have filtered through to the mind--- most definitely not the real reality. The only real reality in us are the natural atoms in our bodies. But once they created us as sentient beings we became foreign to nature unable to observe real reality anymore because unlike natural atoms we can no longer communicate with nature direct.

the natural dimensions of space, or phenomena. Nobody, to be sure, asked the question; Minkowski asked it and proceeded to answer it to the apparent satisfaction of mathematicians. But mathematicians, as people, do not constitute the arbiters of truth. That role is reserved for logic as a subject of which mathematics, also as a subject, is even just a minor part in human thought.[255]

Minkowski just took it upon himself to solve a problem that did not exist, saying if Einstein had paid attention to mathematics when he taught him, he could have made time part of geometry and dispense with the philosophical idea of man rather having to add it to phenomena.[256] As Eddington says, such a system is of great utility. But what of the additional worry it brought about? For, to make time part of geometry, Minkowski had to rely on imaginary quantities. Imaginary quantities just do not exist, or do so only in the imagination of mathematicians; and we know that they are allowed to do that in the course of their work. Yet what was at stake was a reliable concept of the real nature of the external world on which we depend, physically, for our existence. Logically, imaginary quantities or qualities are acceptable in mathematics only when they lead to revelations or hidden truths.

The Minkowski method was declared unacceptable by the philosophers, while mathematicians insisted that it was good enough for them, because it made the new Einstein proposal easy to handle, in the sense that instead of rejecting cosmic time, and inventing our own time before adding it to phenomena, time was supposed to be naturally part of phenomena already---how very convenient. Thence mathematicians and cosmologist only had to mention "The Minkowski Space" (or s=ct...), and all is light, simple. Unfortunately nothing is simple in the study of physical reality. Nothing in nature is deliberately arranged for our convenience. It takes centuries of persistence and perseverance, ingenuity and toil, to discover what we can use to make our lives comfortable. That is the reason we worship inventors and scientific genius.

---

[255] See Russell's essay Retreat from Pythagoras, in his book My Philosophical Development.
[256] It was not a problem in science. It was not bothering anybody, or blocking progress.

I must stress that the whole episode with Minkowski and his supporters in the scientific community, then and now, only goes to justify the novel Einstein proposal of which the debate is all about, namely that in the study of nature time must always be added to phenomena as a separate co-ordinate otherwise theories will be falsified, or be simply inaccurate. That is the logical foundation of the new Einstein interpretation of the external world. The argument over methods of achieving that must not be allowed to obscure the fact that the Einstein original discovery is a vital change in the philosophy of nature and man's relation with it, making man part of the observed. It is necessary to bear this in mind always. In fact, the hassle over methods proves the importance of the discovery, which is that if time is not added as a separate co-ordinate to phenomena, theoretical physics is wrongly conceived, leading to frequent crises and subsequent revolutions. The argument with Minkowski, being universal, is evidence that the scientific community has accepted the Einstein new philosophy of nature as correct. However, if the Minkowski formulation is wrong, then time is not taken into account properly. We have to ponder the consequences of that deficiency in relation to what is now happening in theoretical physics and cosmology, since we cannot get the same results with 4-D geometry as we get with the 3+1 formula.

This debate, however, is the pinnacle of all logical discussions about the nature of physical reality.[257] So, on point of logic, we are obliged to adopt the Einstein proposal. One, because, with his new theory, he was able to solve most, if not all, of the major problems then facing and crippling physics. Secondly, Einstein took account of the role of man in perceiving what there is in the world. Whatever the physical world is, or is not, it is man who apprehends it as so-and-so. This gives man a pivotal role in the determination of physical reality. Man decides on reality through how it is perceived by the human mind---or the brain-eye complex. That is not scientific; it is rather a philosophical enterprise. Yet that is the only way human beings can proceed; our only window on the world; that is what reports of the nature of reality to us; so it is part of the process of determining what the nature of physical reality is, because what

---

[257] The debate is not new. Plato's simile of the cave made the same point.

the mind perceives is all we can experience; all we can have to live by.

I see nothing wrong in man himself adding time to phenomena if that is the only way we can make sense of reality. And Einstein had proved in the special theory of relativity that with time added as a separate co-ordinate to phenomena, several outstanding problems could be resolved scientifically. So the human addition of time to space can be incorporated into science; it does not falsify it; it improves it, as special relativity has demonstrated.

The mind reports on physical reality; what the human mind finds to be there in nature is all we have to work with; so if we add time as a co-ordinate to space and it works, then that procedure is scientifically acceptable. As Kurt Gödel has shown with his theorem, even arithmetic cannot do without undecidable elements. All such propositions are bound to "...contain sentences 'S' such that neither 'S' nor the negation of 'S' can be proved". Thus we are left in no doubt that man cannot know everything; the best we can do is to rely on the data from the senses; whether they are the ultimate truth or not, no one can find out. So when Einstein says we must add what we take to be time to phenomena, that precisely is what we have to do, however the time is defined.

A contrary formula that there is time already in all forms of phenomena and space, must define the time first, since no mathematics so far produced has been able to demonstrate that time is the same thing as space. If not, then we would not know what the theorist is talking about, and must revert to what our senses present to us as 'the time', which at present we know as the periodicities given by a link between the internal sense of duration and external cycles on this earth.

In any case, the two systems (of Minkowski and Einstein) are totally irreconcilable; so we have to choose between the two; and I think that the Einstein theory can be seen as the more attractive and acceptable---because it is not vitiated by arbitrary elements. To logicians and philosophers, the word 'arbitrary' carries an aura of unspeakable horror in matters of serious scholarship; and to be fictitious is equivalent to being fraudulent; yet that is how Professor Eddington described the Minkowski formula.

Thus, since the Minkowski proposal is not acceptable as a valuable contribution to theoretical physics, we are left with the proven theory of Einstein, meaning that, with his mind, his only window on the world, man contributes something to the nature of physical reality. Perhaps we do not perceive the world as it really is, but all we can work with are what the mind reveals to us. And since what the mind reveals is part of the process of learning about the world, man is free to decide how to organise his materials for the control of the harsh aspects of reality out of the data provided by the brain/eye complex we call "mind" for want of a better word.

The Einstein system correctly suggested that in the organisation of the materials found in the world for use in the service of man, adding time to the three natural dimensions of physical reality is the correct procedure---even Minkowski and his followers have accepted this, with the suggestion that the time (is or) can be made part of phenomena with mathematics and does not have to be added as an external factor by man. In other words, Minkowski thought that time is naturally part of geometry, and set about to invoke it with mathematics.[258] As I have conceded, in a way I agree with Minkowski too; I can appreciate why mathematicians adore his system. For if, having accepted the Einstein principle, some mathematical technique can be found showing time to be already, naturally, inherent in space, so much the better.

Unfortunately, what Minkowski suggested can only be achieved with imaginary quantities and therefore, as such, completely unacceptable.[259] But since the Einstein principle has been

---

[258] In fact, time is naturally part of the human mind's *modus operandi*---using points to delimit events---for duration uses points. Whether Minkowski knew it or not, what he did was to invoke the internal time sense with the square root of minus one, or his mathematics. The point is, since time begins in the mind as the sense of duration, any symbol we employ can invoke it; but that is not the whole of time. External cycles are required to have time that can be mechanised for all and sundry in the clock. The external aspects of time make it the product of points as applied to space---or simply the product of space, not part of it.

[259] A proposal to unite space with time should not even be contemplated under relativity, because under relativity the time is derived from the space---how do you unite them again to obtain another concept of time called 'space-time' that is totally different because it is supposed to include space

universally accepted as correct, we have no other option than use the 3+1 formula, whereby man is the one who adds time to the three dimensions of space as a separate co-ordinate to give us a proper understanding of the nature of the physical world.

But the reader should know that everything I have written here is exercise in futility born of my own buffoonery optimism. Mathematicians are presently unwilling even to countenance the mere revision of the Minkowski formula, let alone discarding it. They know it is not a true representation of the physical world, Professor Eddington told them so; yet they want to keep it. One wonders when they will realise that the fictitious 4-D geometry is corrupting all the work they are doing in cosmology and general relativity.

The Encyclopaedia Britannica (Macro) tells us that, "practically all cosmologists assume that space-time is infinite in its timelike directions." The term 'timelike' was introduced by Hermann Minkowski. From what the reader has read so far in this book, I am certain he or she will be able to judge whether time, as space-time, can be infinite---unless we are going back to universal time that permeates all the cosmos like a thread, which is what the Minkowski formula amounts to. For otherwise time cannot be conjured out of thin air with mathematics that rely on periodicities which have come out of the earth-year, as the time for this inertial frame.

Every frame has to have its own time worked out of its own repetitive cycles to get its own 'unique' local periodicities. This is what the Einstein notion of time amounts to; any other practices violate the special theory of relativity and are therefore unacceptable. In any case mathematicians and cosmologists use the Minkowski theory as a formula in their notoriously formulaic suppositions.[260] Thus they would write the

---

by means of mathematics? Is this sensible? Time is derived from space and independent of it so as to have its own identity. You have to have the time in hand before you can unite it with space; but once it is gained, it becomes independent. Logically it can only be added back to space. In no way can it be seen as the same thing as the space that produced it.

[260] "There are two great errors (sic) to which the finest scientific brains are often prone. One is to embrace new thinking so completely that one loses sight of its limits. The other is its mirror image---to hold so inflexibly to the tenets of orthodoxy that one's mind remains closed to new and better ideas."

equation for the Minkowski theory (either the 'ict...' or the 's=ct...') and move on---implying that the 4-D geometry is part of their theories. It is not clear that they rely on the earth-time's unique periodicities in their minds; but if they do, they are violating the rules in the special theory of relativity, which says time is limited to a frame. Our time cannot be applied to other frames without ambiguities, as Russell has stated. No mathematical technique can achieve that feat, for the simple reason that you need periodicities in any time system; and the earth's periodicities are not pertinent elsewhere because they are based on the earth's peculiar postulates, mathematics and parameters.

The nature of other metrics cannot be guaranteed to be the same as that of the earth. It is easy to breach the rules in special relativity, after all Einstein is not holding a stick over our heads; but it should not be done, for Einstein presented his ideas as a logically deductive system. Taking parts away will not do---I mean to say otherwise he could not have solved the problems in physics at the time. Having demonstrated the truth of the special theory in that manner, it would be most unwise to violate any of the rules in his theory without valid logical grounds for doing so. The Minkowski formula cannot provide the valid logical grounds for breaching the sacred canons of special relativity. In fact, the formula is an insult to our intelligence. I fail to understand why Professor Eddington condemned it as arbitrary and yet urged us to employ it only with the weak caveat that it is nevertheless arbitrary and fictitious. What was he thinking of?

No theorist other than Copernicus, Newton and Einstein have been worshipped so religiously for his genius as Minkowski--- yet he had no such genius. Like Einstein, Copernicus was right. Newton was right in many respects. Einstein was an incomparable pure, sweet and lovable genius never before seen on this planet. On the other hand, Minkowski's theory is labelled arbitrary and fictitious by a scholar of the stature and achievements of Professor Sir Arthur Eddington and yet he is worshipped even more than Einstein because it is said that he made relativity easy to understand. Well, if they cannot

---

(The Science Editor, *The Times*, in his Review of scientific Books, Saturday, August 21st 2004---p12.)

understand a minor theory like that they should resign their university positions. Probably they, like Eddington, are worried about what to replace space-time with. If so, they don't have to fret any more. We can retain the term 'space-time', for as Russell has observed (in his *Analysis of Matter*), the concept of merging space with time was already present in special relativity. There is no doubt that the 3+1 formula implies space-time but only in the sense that time cannot be obtained without the use of space.[261]

Thus if scientists cannot live without using the term space-time they can keep it but only in the sense that time can only logically be obtained from space, not that it is the same thing as space. But we cannot go on pretending that a fictitious formula is the basis for defining physical reality---i.e. that because of that formula space is four-dimensional when it is physically not so at all. Of course, saying this will not make me popular and I know it.

Altogether, Minkowski is a disaster for science because he affects cosmology, astronomy and physics adversely. So long as we continue to use what we call time as 'time', almost in the old sense of something permeating the entire cosmos, the Minkowski formula can be used. Yet time is altogether different. We can only know how it is passing through nature and never what it is. We use repetitive cycles to give us the rates of its passage. That is what I call 'Quantifying Time'. With this interpretation of time, 4-D geometry, or four-dimensional space, is not what gives us time. We apply the points to space to give us the time units. Russell's definition of it as 'relation between points' (like how we get the years), is consistent with Quantified Time, not so with four-dimensional space.

But the irony is that (scientist cannot do so, barred by logic), because they can only define four-dimensional space by using

---

[261] Once cosmic time was abolished every system of time became 'space-time' automatically because time requires points and you imply space when you use points. But that means the time is the product of space and points (the years, for instance), a third entity, and therefore in no way identical with space leading to the concept of 4-D geometry. It may be useful to mathematicians to consider time as such. Any such proposal can be used, too, because time is always the same. But it is not physically true of the world and leads to a distortion of physics and cosmology.

time as a coordinate. How do you get the time in the first place before it becomes a coordinate to help establish the 4-D geometry or four-dimensional space of space-time? The answer we are forced to swallow is that it is imagined to be there. There where? By the transformation of coordinates, which is nonsense. At this stage the mathematics covers the four walls, and they urge you to accept that they are right through the use of counterintuitive mathematics.

Because the end results of any formula for acquiring time (on this planet) still gives us the seconds, or the familiar time units, the defect in the Minkowski formula is camouflaged. Why is it that space time gives us the same familiar units of time? It is because the second is the same whether time is four-dimensional or not. Scientists can use this, as Eddington urged them to do, but it is not true of physical reality and I think it is causing distortions in cosmology, astronomy and physics. There is no universal time, but because Einstein used the Minkowski fictitious formula in the field equations of general relativity, and Eddington said it may be used, the latter's caveat against its falsehood is conveniently overlooked. Everybody just mentions 'space-time' as meaning the existence in nature of four-dimensional space. But if so then time is universal again, and the assertion that there is no longer a universal time is false.

What is the real situation? For me the truth is this: not only is there no universal time, but time cannot even be defined and we only use how it is passing by as time without knowing what it is. Obviously that is sufficient to enable us to live comfortable on the planet due to the day and night system, the months and the year. It is easy to plan with these periods. The day and night allows twelve hours each way; the months are long enough; and, of course, the year is there. You can do a lot of planning in one year, knowing that it is bound to come. When we start a new year everybody knows that it would be twelve months before it ends. This is long enough. It still doesn't allow us to know what the nature of time is. But whatever it is, we get no sign of any kind that it is universal to rescue the Minkowski four-dimensional formula.

I suspect the belief in the concept of space-time continuum is induced by mental indolence---people are very worried about the nature of time, especially those inclined to reject religion;

they also confess that it is disconcertingly mysterious. So they stick to space-time, believing that it is scientific---and, above all, even Einstein used it. Is there nobody in the world able to decide that because Einstein used a theory does not mean it is true without investigation, or even when those who have investigated it have declared it to be arbitrary and fictitious? The answer to this question, I think, should be no. Not since the demise of the many great men like Russell, Whitehead, Hilbert, that we luckily had assembled in one century---with Einstein added as a special bonus!

I regard this as my own personal battle with the academia, particularly the mathematicians. It is most unwise to cast aspersions at the academia as a whole because it does indeed produce men of outstanding intellectual abilities. But there is a battle going on between writers of my ilk and the consciences of scientists and mathematicians. For me, it is a matter of conscience because the facts are still as clear as stated about a hundred years ago by Russell and Eddington.

First, let me state the battle grounds---and they are better stated as two questions. Is time the same thing as space, or, in other words, is space four-dimensional? Secondly, is it right to regard time as consisting of only two aspects: the theoretical and the practical where the theoretical is unknown to people outside the field of counterintuitive mathematics, and the practical alone is demonstrable by the clock after the theorists have distilled the abstract nature of time for the ordinary man?

Writers of my ilk argue that indeed there are two aspects of time, but they are rather the physical---being the cycles or orbits of the sun or atomic oscillations---and the psychological, which constitutes the internal sense of time we know as duration by which we can tell the lengths of the different units of time mentally, e.g. that the hour is longer than the minute; so that if we are told to wait for an hour we know that it is longer than waiting for a minute, and so forth. But we cannot agree that there is a theoretical aspect of time known only to advanced mathematicians. Einstein changed time simply by taking the Lorentz 'local time' idea as 'time pure and simple'. The logic of time since then, as discussed earlier on, is that all time is obtained with the application of points to space. Your local space, your local time. That is the reason Russell (who was a great mathematician) defined time as 'relation between points', for that exactly is how we get the earth-year, which is the source of all other units of time as fractions thereof. The year is not conceived by mathematicians in their counterintuitive settings and then translated down to ordinary men and women as 'the year'. We all see it happen. There is no such thing as theoretical time to justify the Minkowski fiction.

In the end there will be no shootout to decide the issue.[262] But since time's interpretation creates civilizations (or mass following to organise life in accordance with 'our time' on earth: e.g. 'Creation' and religion, absolute time and Newtonian science, Minkowski space-time and post relativity science), this ferocious intellectual battle will probably be settled or can be settled only when mathematicians decide, as a matter of conscience, to abandon the Minkowski arbitrary and fictitious equations of space to time. Until then, writers of my ilk will labour in vain---yet still I will soldier on! I am already over seventy and haven't got much to lose. Meanwhile, since the Minkowski formula can be used to justify time travel, several books about time travel are selling like hot cakes. Given that mathematicians, like logicians, are supposed to be consummate intellectuals who pursue truth for a living, one wonders whether because of their mystical streak they are rather enjoying the shift to mysticism inherent in the Minkowski formula.

Let no one deceives himself: mysticism forms the bedrock of all the religions and religion causes wars. To use mathematics (through contortions) to justify such a major shift to mysticism cannot be right. The great Lord Bertrand Russell said coolly that the space-time concept is compounded for the convenience of the mathematician. He means it is not true of the actual physical world. Eddington also said it is arbitrary; what is arbitrary in science is false. I rest my case.

---

[262] This is the sort of issue that used to be settled by just one thinker stepping forward. But the debacle over the eather and the subsequent solution of the propagation of light by relativity so shocked scholars that nowadays they are afraid to advance new ideas. We are often told that even Einstein used the Minkowski formula. But intellectually what does it matter if he did---it can still be wrong, and it is wrong; we are now told that, actually, he never really believed in it. We seem to have reached a stage where scholars and thinkers are not working so hard any more, but prefer to mention authorities for the truth—a well-known religious practice!

As I have said repeatedly, I am very old and far gone beyond worldly success so I can afford to say these things without fear. Many others feel the constraint of organisational pressure. So I am appealing to their conscience. That is all anybody can do. That I am everywhere ignored is of no consequences to me whatsoever. I am so old and weak that I couldn't care less. And I have churned out ten 'defective' or imperfect books all saying the same thing. At least one of them will survive. All scientists must be brave enough to agree with Sir Arthur Eddington that the Minkowski equation of space to time is arbitrary and fictitious and therefore false---time is time and space is space, except that, according to Einstein, you cannot get time without the application of points to space, and since that precisely is how we get the years, Einstein is right.

18

WHAT IS ATOMIC TIME

Atomic time uses the pulses or oscillations of atoms to show how much time is passing. The passage of time is important because we count these units as they occur, not before they occur. Yet, as units, and being successive, they are gone as they occur: thus time is always passing, whether as orbits of the sun or oscillations of atoms. So long as we can only reckon time with successive cycles, we can only tell how it is passing and never what it is. Ordinary earth-time uses the orbits of the sun as pared down to the seconds. Technically, both earth time and atomic time are using regular motions or cycles to show how time is passing. The only difference is the comparative accuracies of the techniques.

Many scientists (probably closet religious believers) often refer to atomic time as a different kind of time from earth-time, and that it is evidence that time is super-naturally there in nature and man measures his version of it; they claim we can do that with even single atoms, or anything suitable for the purpose---which is the religious view of time as opposed to secular time; because in secular time we suppose that we do not even know the nature of time only how it is passing by through the use of regular cycles; and that our units of time are the repetitive cycles---the years, for instance, as pared down to the other familiar units, beginning with the second.

Obviously, sentience is required because somebody must be there to count the orbits of the sun as 'years' or there will be no years or seconds based on the year.[263] Theoretically, this secular time is defined by Professor A. N. Whitehead as "A time system is a sequence of non-interacting moments [that is, year after year after year, or second, second, second...]"[264] So the question is this: is atomic time different from time by the motions of the earth round the sun, such that it casts doubt on our definition of time as the rates of cyclical motions (like the years), used to show how time is passing through nature and never the true nature of time? The same question may be put differently, namely is there a natural time that even single atoms cam be used to invoke? My answer is yes, but, like the earth's orbits, only how that time is passing by.

The radiation pulses of caesium 137 are used to measure the length of the second more precisely.[265] That is what we call atomic time. However, it must be noted that it does not mean atomic time measures time from nature more accurately than earth time; rather it measures the second more precisely in relation to the mathematical divisions of the length of the earth-

---

[263] Anybody who is not a philosopher is cowed and becomes a cowardly believer when he or she looks too deeply into nature. This is one of the mysteries the religions exploit to remain powerful over mankind. Time, they think, is so strange and impossible to understand that it must have been created by God.

[264] *The Principle of Relativity*, Ch. IV.

[265] The deliberations at the United Nations' International Telecommunication Union about world time, time zones and the atomic time cannot vitiate the theory of time presented in this book. Atomic time, in particular, is not different from the time we get from the orbits of the sun. The oscillations of the atoms are only smaller, but they have to be regular to be used as time---in which case they can only show how time is passing and never what it really is. As I have said, any regular motion can be used to show how time is passing. You can even tap the finger for the same purposes. We can never know the true nature of time, but. to keep track of how it is passing any repetitive motion can be used. We like to make mystery of time. We are right; it is strange. The truth is that we do not need to know its true nature; knowing how it is passing is all that we use or need to use as time on this planet.

year. As such, the atomic time oscillations are not different, logically, from the regular cycles we use for reckoning traditional time. They merely consist of shorter cycles, showing the rates of the passage of time like any other system we can imagine.

Secondly, the atomic time is useful only as related to the second, which is one of the fractions of the earth-year; so it means we are still using the earth-year when we rely on atomic time. The atomic oscillations can never tell us the true nature of time; they are the same as the long orbits of the sun. They show how time is passing by only, except that their pulses or oscillations are shorter. Actually, atomic time confirms the nature of the entity we call time. Atomic time is passing by; the pulses or oscillations of atoms that we use for time are passing; they do not persist; they do not last; they do nothing but just pass by. If you need ten seconds to boil an egg, the oscillations of atomic time will require millions, simply because they are so short. But the principle is the same---like earth time, they are passing by. Only when the required oscillations have passed and been counted as accurate will your egg be done. So, as time or an indication of time, we know only how they are passing by and never what they really are in nature, although I guess they are related to chemistry and motion. However, as discussed in 'Process and Reality' in this book, The chemistry is a 'process' in nature; to man, the reality of that process or event is what we experience as time or a period of waiting.

CONCLUSION

Since the Introduction is rather long, this epilogue is kept short. What I can promise the reader is that it is written in the style of a literary yarn or a story, eschewing all technicalities or most of them! I believe the best scientific theories are those that can be presented as stories so that the lay reader can understand them. As against this attitude, the Cambridge University Press have rejected the book partly because, "…in its style the work is addressed to a relatively general audience, which is a readership that our philosophy list does not serve".[266] I know what they mean, but if I took that route I would have to present my ideas with strict reference to the academic papers and books by professors about time, most of which are so opaque and inconclusive that only other professors can understand them, and the lay reader is effectively excluded.

---

[266] Their other objection was that the book (then only 58,000 words) was too short. But that was not a serious objection because they could have asked me to add more materials as I have done. But the objection to the general audience annoyed me most, since I see it as a class matter. Why don't they serve us too? Are we not good enough for the CUP alone because it is the best in the world? More than a thousand publishers will dispute any such arrogant claims.

Yet about time alone of all subjects, I take the view that everybody is involved. It is the commonest thing available to everybody, and it is not fair to deliberately exclude ordinary (intelligent) readers who may have something more interesting to contribute to the debate than the opaque ivory tower arguments from the complacent dons of universities. Cambridge made Wittgenstein professor of philosophy to replace G.E. Moore at a time when he was writing what could put an end to physics, as Bertrand Russell has related. I wonder what the scientists at the Cavendish Laboratory think about that. Many Scientists at the Laboratory have won the Nobel Prize for physics, and my question is, would the Laboratory appoint Wittgenstein professor for any subject on this planet at all, let alone as professor of a philosophy that puts an end to physics? Of course not. No theory can do so; therefore it means Wittgenstein was talking nonsense and yet Cambridge made him professor. That says a lot about Cambridge.

I think the CUP is mistaken in elevating their publishing house even above the status of CERN or the Cavendish Laboratory where only specially trained scientists can contribute to their work. Even then these major (and certainly unique) institutions do communicate lucidly with the general public; they want us to know and appreciate what they do with the taxpayer's money. Why is the CUP (a mere publisher and certainly much less important than the OUP) so smug? I think this is a class matter. Given that Cambridge is often accused of snobbery, to state in print that they do not serve the general public (whatever may be the subject) is a serious insult to the taxpayer. For even Einstein and Bertrand Russell deliberately communicated their extremely complex ideas to the general public—and here we are talking rather about a subject (time) that is known and used by everybody. I don't mind being rejected, but not on such arrogant, pompous and snobbish grounds by a public institution.

In the end I blame myself for making any kind of approach to the CUP about time, which, in their ignorance, they regard as part of philosophy. In fact, there are two aspects of time in the human mind: the repetitive, physical motions used to show the rates of passing time (the earth's orbits, for instance), and the mental periods, or duration of units of time deliberately assigned by us to represent the mathematical units in the mind so as to be able to tell that one minute is shorter than one hour. But, considered as philosophy, perhaps I was literally insane to approach CUP. Apart from the mad thinker, Wittgenstein being honoured with a professorship at Cambridge, CUP published Professor Kneal's book "On Having Mind" as recently as 1962, in which he praised the Platonic myth as against physical reality such as we have established with science. Is that not another example of bypassing science about which the Cavendish Laboratory can do nothing? I think C.P Snow's "Two Cultures" did not go far enough.

The Introduction and Conclusion of any book serve the same purpose, namely to inform the reader about what is in the book. What may strike the reader as deviations are all part of the discussions of time, for time alone affects everybody and everything we do. The only truly universal subject in the world is time; the irony is that it does not come from the universe but created (in the words of Russell, 'constructed'), by human beings, which is what the whole book is about.[267] Russell did not say this lightly. It is meant to presage a new philosophy of time and human existence, since everything depends on time and the nature of time. Any change leads to a whole new conception of all existence including the universe itself. We apply time to the universe, trying to know how long it's been there. That is the essence of the problem of time, namely how to know how long. We cannot say the universe has always been there. According to general relativity it is changing constantly---but since when? What is the time we can use to determine that if it is true that we construct our own time? After all these pages of arguments I still don't know it. Nobody does, and no matter what we do, I am convinced that nobody can ever know that. People should get used to the idea that man's brain is puny and not capable of answering such questions in a billion years---of our time![268] The point is, in the absence of a

---

[267] It should be a small book, as my other books are mere monographs. What makes this a medium-size book is that I decided to say something about the role of religion and the Minkowski formula almost on every page. One, the religion, promotes myths and illogical ideas about time; and the other, the Minkowski equation of space to time, is logically untenable; and I had to condemn them at every turn because they undermine rational theories of time.

universal time, what, indeed, is our time? How do we get it, and how long (or how far) does it go into the nature of existence overall? Another question is this: what is the value of the astronomical light-year since the year is our own unit of time, yet astronomically the sun is so tiny that orbiting it may not mean much in terms of duration or distance if we ask how long is one year? A lot of time has been wasted on the Minkowski fiction instead of finding answers to many relevant issues.

The Introduction is a projection; the Conclusion is a kind of summary, and may be slightly different from the Introduction because it is what has actually been put down; for in writing a book many ideas occur to the writer that he or she could not even have thought of during the planning of the book.[269] One of the joys of writing is to invent ideas and expressions as one goes along, and at the end ask oneself, "Did I really write that?"[270] when the beauty of the text surprises him or her. Like the best poets, when you write a brilliant phrase, it hits you with an intoxicating thrill a like a bomb that invades your body excitingly. One does not always know what one will eventually write down, or remember all that he or she has written. For instance, Bertrand Russell's *History of Western Philosophy* is so detailed and brilliantly written that he certainly could not have recalled all that he had written.

So much has been said about life that by now I believe it is generally accepted that life was a chemical accident. Russell said it was an accidental collocation of atoms that knew not what they were creating! But we are reasonably certain that nobody sat down to plan human life; he or she could have made a better job of it.[271] Time which is very closely associated

---

[268] Without literature, recorded history, the level we have attained in our scientific knowledge, social customs and all the products of human intelligence, no one human being (however clever) is worth much. You can't do much on your own---you need the accumulated knowledge from the earliest times to the present day behind you.

[269] The book is entirely about time: what is the contraption we call time since it does not exist out there physically, how did it begin, where does it come from---God or the cosmos---and how it passes by and seems continuous.

[270] Another thing is that no one ever publishes all that he writes down. You cannot even type all you have written by hand. This is the reason researchers like to dig into manuscripts for nuggets of wisdom missed or ignored by the printer.

with life (at the very least because we age 'over time' and die through ageing and therefore killed by time if we are lucky enough to escape the murderers), has been mysterious from the beginning of life on this earth. We cannot do anything without it. But can't even know what it is, except that it is always there and moving on to make us age continually. Nothing in life is more mysterious other than why the life itself was created through evolution, not Creation. This is the reason discussions of time include matters of life as well.

When time is logically or scientifically scrutinised carefully we find that, according to H. A. Lorentz, it varies from place to place, so that he called each time 'local time'. Or, to put it conventionally, he found that, contrary to the notion of absolute time, time does differ from place to place. This was an actual, physical discovery; and the greatest logical thinker in history seized on it to give us 'The Einstein notion of Time' based on dynamic space and therefore, essentially, dynamic time. Nevertheless, I have stressed in this book that, since not all of us can go into the theory of dynamic space as presented in relativity, the term 'changeable space' will do. That is to say, time is changeable; which is still revolutionary because it was formerly regarded as fixed or absolute and generally covering the whole universe and the same everywhere. Before Einstein nobody could imagine that time, as mysterious as it is, was not divine, much less as limited to a frame.

Through the analysis of Order and Simultaneity, Einstein recommended that the Lorentz concept of 'local time' may be called 'time, pure and simple'. He, of course, did not put it so succinctly, but that is what he meant, since he went on to state that "There are as many times as there are inertial bodies." In plain language, every 'body' will have to create its own time because there is no such thing as general or absolute time such that a second here is the same everywhere else---, in other words, time is your own local time. As a result we have come to realise that time is neither divine nor cosmic. Bertrand Russell was, by proper academic estimation world-wide, the greatest philosopher of the time, logician, scholar of the highest

---

[271] Religion had nothing to do with the origin of life on this planet, for after all the religions were invented after man arrived on the scene, and they are clearly the work of our fellow men with designs for human domination.

rank, mathematician and philosopher of science who actually invented (or laid the foundation for) the philosophy of science with his book "Our Knowledge of the External World". He thus concluded that time, technically, is "relation between points"--- such as the manner in which we get the year: from one point to another, and further subdivide the year into fractions with mathematics or points, so that a certain number of seconds, hours and days will exactly equal one year and then we start another year. That is all human time consists of. So there is no longer a universal time. The problem of time was seen as searching for the origins of our time, how it passes by and seems continuous and, most important of all, what it is.

The definition of time as the irreversible general passage of existence was meant to refer to general time. Once Einstein discovered that time is not general or absolute, that definition became untenable. For if each inertial body is to have its own time then time is bound to differ from place to place and cannot be 'the general passage of existence'.

Thus began the idea that there is such a situation as time before Einstein and time after Einstein, as discussed above. The contribution of Lorentz in all this was overlooked! Time before Einstein was general and absolute (some call it Newtonian absolute time) so that an hour or any unit of time here is the same everywhere else. Time after Einstein is not general but limited to a frame. It is also not absolute but dynamic. A few dissenting voices were silenced by Professor Eddington for making "meaningless noises", as quoted above, in his great book, "The Mathematical Theory of Relativity"; for he was the leader of the scientific team that confirmed the truth of the general theory of relativity, known as the bending of light in a gravitational field.[272]

While Eddington hailed the Einstein notion of time, Einstein was also singing the praises of the Minkowski notion of 4-D

---

[272] One small criticism of Professor Eddington is that it is in this same book that he described the Minkowski formula as arbitrary and fictitious, yet he added that "...we shall continue to employ it..." I can only assume that, for mathematicians, it was easier (at the time and probably still now) to use the concept of 4-D geometry for their own peculiar interpretations of relativity than the 3+1 system.

geometry, or four-dimensional space that made time part of space with mathematics alone. But for Einstein, after a brief sketch of the formula, he declared: "These inadequate remarks can give the reader only a vague notion of the important idea contributed by Minkowski. Without it the general theory of relativity...would perhaps have got no farther than its long clothes.[273] Now we learn that he never actually understood the formula. We need to search for the truth about this matter.

Whatever may be his real motive for doing so, I found it strange that Einstein praised Minkowski because his own notion of time was a contradiction of the Minkowski formula. The Einstein supposition means time is invented or created by individual inertial bodies; so every 'body' has to have its own time. On the contrary, the Minkowski formula makes every space "space-time", the same thing as time. This means time is in every space; it is not created by individual bodies. It is there already. The two theories cannot be reconciled.

Furthermore, the only method for creating time for individual bodies is the orbits of the sun, i.e. by repetitive cycles. Either orbiting the sun or counting the pulses of atoms. It is the same thing. But that gives only discrete time, not one that is present in every space. Under the Einstein proposal we are quite happy with this because time is generally known and used only in units.

I have demonstrated in this book how happy I am to note the report from David Hilbert, quoted in Professor Yourgrau's book, that in fact, Einstein understood less about 4-D geometry than schoolboys on the street. What that means to me is that he never truly accepted the Minkowski theory of four-dimensional space, or space-time, as logically valid.

Another defence is that at this time Einstein and his theories were subjected to intense scrutiny. One of the recurrent criticisms was that, with sufficient mathematical acumen, he could have formulated his relativity to dispense with the unscientific 3+1 system and incorporate time into space. And that precisely was what Minkowski was offering. So Einstein

---

[273] From his book "RELATIVITY", Opp. Cit. Sect. 17, Part I. Perhaps he was speaking to his mathematical controllers who were urging him to use the Minkowski theory, because the Minkowski idea (or 4-D geometry) was in no way relevant to the central thesis of general relativity.

was eventually persuaded to incorporate the Minkowski formula into his general relativity. Unfortunately, the basic logical premise of the Minkowski theory is known as his ict equation, upon which it is deduced that s=ct.[274] Thus, according to Einstein, and in his own words, already quoted above: "...we must replace the usual time coordinate t by an imaginary magnitude $\sqrt{-1}$.ct proportional to it." Bertrand Russell did not like that very much and said so plainly in Chapter XXXVIII of his book "The Analysis of Matter".

In spite of all this, the theory of time outlined in this book claims that the nature of time is unknown; but I agree that Einstein came closest to what it is. It is not his fault that every method we can use for keeping track of time is only capable of showing us how it is passing through nature and never what it is. It is conceded that time is closely associated with chemical processings in nature---physical or organic---but we can only know how it is passing by using repetitive cycles to give the rates of its passage.

There is no need to go into the complexity of Minkowski's mathematics. Of course a book on physics would need to do that; but this book is entirely devoted to time. It is not even an elementary book about physics to discuss the Minkowski formula in outline. Since we all agree that its basic premise, being his ict equation as quoted from Einstein himself, is logically flawed, his whole system must be rejected. Reproducing his mathematics (which means just copying it out) would not save his formula. What amazes me is that everywhere it is described as artificial or arbitrary, and yet all scientists refer to space as 'space-time', the same thing as 4-D geometry or four-dimensional space, which is what the Minkowski formula is called. He's had this strange success because physics is easy to twist by means of mathematics. Mathematicians reason in formulaic modes. So when something goes wrong, as happened with the eather debacle before Einstein, everybody is lost because everybody is using

---

[274] All this has been explained above. It means because ict is true s=ct (that space equals to time), where s represents space and ct (together) is for time as observed by light. This, in full, was the basic equation for space-time or four-dimensional space, making space and time into one entity in mathematics, not in actual physical reality. Yet mathematicians and cosmologists continue to use it. I call that a distortion of physics.

the same formulas. The fact that Einstein used the Minkowski formula without believing in it, should warn scholars that sometimes some formulas contribute nothing to a thought process, and sometimes, like the example with the eather, they can evidently lead to or create distortions that may take years and the efforts of a genius like Einstein to resolve.

Even though this is just a book about time, because of the Einstein example (of combining philosophy with physics and cosmology), I have been forced to discuss the possibility of training Philosopher/Scientists in the future. One reason is that a book about time is not just a book about the clock, but is bound to involve questions concerning existence since time and existence seem to be very closely connected. The samples of texts from scientists and philosophers make some philosophers seem to be toying with science. Yet, as Russell has warned, we have to obey physics on pain of death. Nobody will ever forget Hiroshima.

The ITU is often mentioned as the relevant authority about time on earth. It refers to the United Nations' International Telecommunications Union that settles disputes about International Datelines. That organisation has nothing to do with time in metaphysics. It merely decides zonal time systems and legal and practical accuracy of clock systems.[275] What is more important is the questions people ask about past, present and future. This common syndrome has been explained in the book as part of the myths and legends of time. It has absolutely no physical basis other than the fact that it is related to the day and night system. It leads people to think wrongly that past events that belong to yesterday and beyond are actively still going on in the past; they forget that the events and the people involved at the time have moved on to become the present and going on to become the future, too.[276] It's all a myth that they are still existing in the past as they were experienced. Yet,

---

[275] The important question is that asked by Bertrand Russell, namely if cosmic time is abandoned, what is measured by the clock? United Nations has its dirty and incompetent fingers in every pie but the ITU cannot answer that question. I hope my opinions about that are sufficiently cogent.

[276] It is a surprise to me that people do not realise that today's events are rather the continuation or the consequences of past events, which therefore are not going on in the past but rather going on right now as part of the continuing story of human existence.

strangely, people have worked past, present and future into an extremely complex maze of religious sentiments, scientific theories, mathematics and some facts of life, including imaginative ideas about the possibility of travelling back to revisit past events or forward to meet the future.

Time in the clock is quantified time; it is obtained from repetitive cycles, the years for instance, or even the pulses of atoms. Future events have not yet occurred, and past events are not still going on behind us because it is what we are presently dealing with: present events are those of the past carried forward. So if they continue to exist at all they are here with us! Your troubles are never left behind; they are with you still or its consequences are; similarly, your successes are what you carry forward with you to make you still famous.

Quantified time is discrete since we obtain it with points. Discrete time cannot go forward or backwards. It is finished (like the end of the year) when its unit is expended; that is the reason we have to start another year on 31st December every year.

This is all I know of time. Nobody knows what it really is. From all considerations, what we call time (the years or seconds, for example) is the manner in which it is passing away.[277] And the passage is in units because we use repetitive cycles (continuously) to establish the manner it is moving through nature, or passing by: year after year after year all the way to the centuries; otherwise there are no years in nature---there is only one year repeated over and over indefinitely. In fact, it may not even go on indefinitely because we know that life on this planet is due to be extinguished in four or five billion years time when the sun dies. Above all, and I repeat, the past is not existing anywhere to be revisited. It is here with us. The past is the inevitable baggage (social, psychological, physical---everything) that we carry with us as the effects of past events. So travelling backwards to deal further with past events is nothing but sheer fantasy. **Human beings are peculiarly prone to frequent attacks of senseless fantasy because of dreams and the imagination.** The brain invites us to struggle to achieve the impossible as tantalisingly displayed in dreams.

---

[277] I repeat that the arrow of time notion is no longer needed; it could not logically be defined anyway.

You dream of winning the lotto and squander your income on it without any guarantee of success!

About history, only events have causes going as far back as possible, and also have effects going forward forever as the continuing story of human life. So history is not the march of time; it is rather the march of events. The dates are merely added as the time, day and year of occurrence; neither is there any mystery in the fact that we have memories of the past and knowledge of the present while we use the imagination to figure out what the future is going to be like.

Human beings have done very well for themselves. We started as mere helpless orphan/apes that ate one another for dinner.[278] Later we stopped roaming the bushes for food and settled in villages and hamlets, eventually building towns and cities; we learnt agriculture to grow food and stopped the practice of hunting and gathering food from the bush. We also learnt how to domesticate some animals. Soon we were building schools and churches (all these, of course, were due to the efforts of the cleverest among us not to the dumb criminal class, paedophile priests and drug peddlers.)

It was not long before we began to look into the stars, build universities and establish institutions to promote learning. Eventually science arrived, generally spreading throughout the population, and not long after that we were reaching to the moon and beyond, creating some forms of life in the laboratory and thinking of settling excess populations on other planets, because soon we will have multiplied ourselves almost to the extinction of life on this planet, thanks to the strong, vibrant and intelligent sperms we derive from scientific agriculture! The result is that we have become vain, arrogant and ready to challenge the Gods; worse than these evils are avarice and

---

[278] It should come as no surprise to anybody that modern man is as savage as ancient man, because if they look into their own souls, they should agree with the philosophers that man has not improved either in character or behaviour, except that the law is more effectively and widely applied to enforce good behaviour. That alone has been more successful than earlier times. But essentially we are all just as greedy, selfish, religious (out of the fear of death) and ignorant of life. A single example will do: castaways always kill and eat their fellow men as people have done all through the ages. What have definitely improved permanently are the laws and the social, economic and political systems, thanks to the liberal thinkers.

deliberate instigation of strife in mineral-rich countries where millions are dying needlessly, especially in the primitive regions of Africa, thus giving the religions cause for concern; yet they cannot help because they no longer have any authority since God is said to be dead.

The eternal curse of man is that the more we progress in knowledge the greater our abhorrence of myths and the need to control evil, yet the religions need their myths. That leaves us exposed to the crazy fools who would rather destroy the world in the name of mythical Gods than live normal lives in which they count for nothing. But who would vote such mad men into power? Besides, not only the vagrants are bent on destruction; the human mind is so delicately balanced between good and evil that we simply cannot know who will turn out to be a terrorist and who will not. Even churchmen have been known to commit acts of terrorism. Training and academic achievements offer no guide of any kind whatsoever. The best trained people are often the worst offenders when they decide to be evil. Money does not help; all despots steal and bank stolen money for their safe financial future and continue to butcher people nonetheless. Sometimes it appears that the number of people killed is more important to despots than the millions made---especially in sectarian conflicts. Altogether, how, when, and if the evil in men will be aroused have absolutely nothing to do with their wealth or knowledge. Some people just wake up one day and decide to die, taking other innocent people with them. God was our universal salvation until the evil among us realised through the excesses of religious leaders that God was invented to keep them in their places.

So, sadly, we have not grown kinder with increasing scientific and philosophic knowledge derived (particularly recently) from Russell and Einstein.[279] Instead we think Hell is other people, a

---

[279] Einstein of course helped with the Manhattan Project (to end the war, frustrate the Nazis etc.), and after his death Russell, too, was shown in a brief film saying many good things, one of which sticks in my mind to this day: "Remember your humanity and forget the rest." I think he was right; they are well-chosen words because avarice is spreading, especially among the very rich, but avarice is always contemptuous of humanity. I believe that avarice devoid of humanity is just about the worst evil on earth---and it's increasing especially among the very rich!

philosophy that, I think, is partly responsible for some of the evil acts of people in the world---instead of loving their neighbours as themselves. But running to other planets will not save us. We will simply carry our bad habits there.

However, intellectually, we can say we are reasonably certain that life was not planned by anybody---the Aristotelian 'Third Man' argument is enough. Besides, we know what atoms can do as we can create other forms of life in the laboratory. We can cure the body of many ailments through chemistry. We can create things like food by agriculture, also medicine and gadgets through the manipulation of atoms.

And now we can also reasonably guess the nature of time and how it passes by. These two conundrums, life and time, were traditionally the only things on earth attributed to God because nobody could imagine how they came to be in existence.[280] Time in particular has become the last hiding place of God since the rise of science or since the influence of Charles Darwin. But thanks to Einstein and Lorentz we have now got a fair idea of time. The so-called 'great religions' claim that they helped mankind to survive during the ancient and primitive dark ages when human sacrifice was the common practice. True or false, we must now thank them and point out that they have also caused wars through sectarian strives and that enough is enough.

Of course, this whole book is about how Einstein solved the problem of time; and very often people who do not realise I was not even fit to wipe his boots ask me whether I think my hero will eventually be superseded as even the apparently invincible Newton has been cast aside because of Einstein. Well, whoever solved the problem of time is the greatest thinker in human history; and whoever solved the problem of life is God. It is possible that many of his theories will see changes in content; but a few of them (like the bending of light in a gravitational field, the curvature of space causing gravity, the propagation of light without the eather, and the quantum theory), seem to be fundamental to the basic set-up of the universe so much so that one cannot see how they can be completely falsified. There could be changes; but the train of

---

[280] Life is certainly a chemical mistake, but what about thought, time and death?

his logical thinking seems so accurate that some of his ideas will always be relevant. So long as the universe around us remains as it is, it is safe to predict that Einstein will remain the King of logical thought on this planet.[281]

Finally, given all that has been said in this book about time, the phrase "The Beginning of Time" or "The End of Time", may be popular but meaningless. The end of any period of time is the total expenditure of its units. The year is ended on 31st December. It is virtually the end of earth time, except that another time will begin instantly---that is the beginning of our time after it has ended at its last point. This is the reason we can plan with the coming year or twelve whole months. We do the same with the twelve hours each way of the earth's rotation. The year comes to pass from one point to another. At the last point, it is gone and we start another year. What people want to refer to is "The Beginning of Life or Existence, and "The End of Life". The traditional ways of expressing them imply a universal time, which, as we have seen, does not exist.

One general error about the interpretation of time is to use the many applications of time to interpret it philosophically. It is not really an error; it is the normal thing we do, all of us do, since we know no other way to describe time. Despite our many means of communication (mathematics, sign language, words, looks, music, gestures and so forth), human beings cannot describe time other than how it is used, or the application of time. Suppose you are asked to tell when you can start work in a new job that you are happy to get, and you mention time---the time you can start, and you are (perhaps oddly) asked to explain what you mean even as a joke. To be safe you can use the law and point to the clock. What the law accepts as time (as expressed by a clock) is time and that is that. Yet, philosophically, the clock is an application of time---one of the

---

[281] Many people are convinced that, since Einstein is only human, he will eventually become as out-of-date as Newton. Yet Newton is not completely out-of-date. Not all of his ideas have been proved wrong. The difference between Newton and Einstein is God. Newton said God made the elements and set them in motion. But Einstein would investigate the secular or logical causes of the elements---hence he traced gravity to its physical causes; and further traced the atoms to the quanta---the smallest bits of matter that can exist. Newton originally discovered gravity, but he did not know how it was caused, and did not try to find out.

many ways of displaying time which has been discovered elsewhere; so it is applying what has already been discovered. The clock, again, is merely reproducing units of time deliberately programmed into it for reproduction. It is not creating time originally for us to use.

The truth is that in life we have only two ways of living, or two phenomena for organising our lives logically in consonance with reality as spread out before us. They are simply objects and how we use them.[282] Science is an additional bonus that makes trial and error sometimes unnecessary either because we can chemically create things, or know from theory what we can use them for. In this sense, what is the object we call time---we know the many ways for the application of time, but objectively what is it? Every mention of the word 'time' in an attempt to answer this question amounts to 'The Application of Time', which means using the application of time to define time. Logically we cannot employ anything's usage to define its essence. Yet whenever you mention time and want to use words (or whatever) to show what you mean by it as a means of definition, you will find that the end result on analysis is application of time---one of the many applications of time. To show time as an entity distinct from its application is impossible except in philosophical of mathematical theories. The application of time shows how the time obtained from the breakdown of the earth year is used, or has been set in mechanical motion to coincide exactly with the orbits of the sun. So using the clock is "application", not the creation, of time.

To explain anything by how it is used is as futile as trying to explain how the motor car came to be invented by demonstrating that it is comfortable to travel in a car. For example, the notion that time can only be known by how it is passing by is often criticised with instances of the use of time for forward planning (this is going back to how we use time for planning purpose as explained above.) Yet that shows classic philosophic ignorance. For we know that the day and night system have twelve-hour periods in each phase. You can do a

---

[282] Originally human beings lived blindly; there was no guide from any source. We imagine that thousands or even millions have died from eating poisonous fruits (as still happens in some countries, especially around the Amazonian forest.)

lot of planning with a gap of twelve hours. (Similarly with the year or any other unit of time, if it is long enough.) But what we want to know is how these twelve hours (or other units of time) came to be in existence, or became available for cultural use. That could only happen with the use of repetitive cycles which are counted as the rates of passing time; and, as such, can only show how time is passing by.

It means time does exist only we cannot find what it is; if it did not exist repeating the yearly cycle would not mean so much to us. For it is a mere physical journey. But it shows how this mysterious time is passing by because it makes us age. You are never the same when ten years have passed.

It should be noted that a second is part of the yearly cycle; it is not a cycle on its own since it would not exists if the year did not---the same thing applies to all the other fractions of the year. The term 'passing time' reflects the fact that you can only count a cycle after it has occurred. But the year and some of its fractions are long enough for us to use for planning purposes. Hence we know that the day time will last for twelve hours and plan how to utilise them. The same thing applies to the night time, too.

In the end, I feel I need to apologise to the reader for the inability to write a beautifully crisp and smooth book; but people should know that the smooth academic books have taken years to develop, sometimes through initial criticisms, and sometimes by several scholars and editors. Original books are almost always messy. For in writing this book, which is deliberately aimed at suggesting a change in attitude to what is called time, I have had to repeat myself for emphasis or through my own errors; sometimes examples are repeated by mistake. These minor faults, surely, are forgivable as a small price for the truth; and whoever is seeking truth, whether right or wrong, deserves a little charitable consideration.

My confessions above of my own personal inability to write out my ideas as brilliantly as Bertrand Russell could do, is certainly no excuse for sloppy writing. I am not proud of it; I can only plead that it is the best I can do. However, I want the reader to be a little charitable, or kind, to original writers. It is their own self-inflicted torture; nobody is inviting them to write speculative ideas that may turn out to be false and only expose them to mockery. Yet they do; and some of them achieve success to help human life on earth. It is nevertheless a very painful process. Any attempt to solve a major problem (be it scientific, logical or philosophical), is always fraught with difficulties, and can never be as smooth to the struggling thinker as that of an established scholar can be; for the latter may have had years to perfect his ideas or thesis. The main reason is that the original thinker is almost always wrestling with concepts that seem to be going away from him---tenuous, ephemeral and ghostly. For these ideas are not printed anywhere for him to copy. He rather have to work hard (99% perspiration, 1% inspiration)[283], to write them down as they occur and recede in rapid succession, and even then he or she is obliged to present them in a specific form to aid the reader's understanding, no mean task.

Let me end by stating what I consider to be the truth about time and the excuse for writing the book, however badly it is written: everybody thinks we know what time is. But rightly or wrongly I have found that, on the contrary, whatever means we use to reckon time, we can merely know how it is passing only. The best example is atomic time. We count the pulses as the rates of passing time---but that is what is meant by (atomic) time. We only know how the time is passing after the pulses have occurred and been counted not before. In my arguments, I think I have shown that, (a) there is a universal logic or method for creating time in the absence of a universal time that was assumed to permeate the entire cosmos and further said to have been imposed from above (fixed and absolute), which makes an hour here the same everywhere else, and probably divine in origin; (b) The old system of writing history as 'The March of Time' is wrong, because the way we get our time makes it necessarily discrete. Discrete time cannot march.

Unquestionably, in my opinion, history is the march of events, which alone can have antecedents and consequences---causes and effects as the continuing story of human existence; also discrete time rules out the possibility of time travel forwards or backwards; even then we have to know what time

---

[283] Great insights, discoveries and inventions are not gained through conscious thought alone; most of the time they appear out of the blue in a flash from the unconscious to the conscious mind. The important thing is the input and the efficiency of the brain sometimes called genius, eureka moments or brain waves.

is to be able to travel by it, but we don't; (c) there is no need for any theory to explain how time is passing. For what we call time is how it is passing only; (d) the idea that time is the same thing as space is rejected because it is only possible to achieve the 4-D geometry on paper by mathematics and not in physical reality---even then such mathematics has to be based on imaginary time coordinates, which is logically untenable. Therefore there is no such thing as four-dimensional space, or 4-D geometry, and the works of many cosmologist, astronomers and physicists that have been influenced by it have all been in vain.

A little book putting forward these fundamental ideas is most unlikely to be masterly and crisply written like an ordinary academic book, and I hope the reader will charitably pardon my mistakes. In my judgement the importance of knowing more about time as a secular entity is to help us realise who or what we are in nature---absolutely nothing. In size worthless; in value none, nature has no values; it is always a blind force. But that should be a blessing. At least we know nature does not value us, but we have values for ourselves; so we should club together and help one another, enjoying the little we can wrest from heartless nature.

To see the logic of this idea my best advice is to read The Human Situation by Professor Dixon mentioned above---it's a thousand times more valuable than any holy book in existence---topped with everything written by Bertrand Russell in literature only. There are other things in art and science to consider outside literature, always guided by the human situation--- so as to 'remember your humanity and forget the rest', that's what Russell told us on his deathbed. Most of the religions exploited the fear of the mystery of time to gain support; but the religions incited wars among peoples. Now that we know what it is we call time, and since it is entirely secular and logically understood, there is no need to support any religion for slaughtering other people out of the ignorance of the nature of time. The fact that there is time and it is mysterious doesn't mean there is God or that those who do not believe in your God should be slaughtered---in any case, the time can now be explained in simple secular terms. That is what I mean. Or to justify my method of writing for everybody and not only for the ivory tower dons (about time alone), let me put what I mean in plain language. If there is no longer a universal time under relativity, then until you create or invent your time you will have no time---and we know that far back in the past primitive man had no time but marked the passing days and nights on walls to indicate the passage of time. Now that we have time theoretically covering the whole planet, it is necessary to probe logically as to what it is we call time or why we have this time; and our search has uncovered the truth that we only use repetitive cycles and count them as the rates of the passage of time---still without knowing what time is.

I think time, or the sense of waiting, is written into the human gene because it originally took 'time' for the basic human gene to form in the cosmos. The result is that every 'thing' requires its 'time' for its existence. This is not the same things as the 'dawn of time'. It is different. There is no dawn of time but the dawn of existence due to time, that is after a period of waiting---time is eternal and nobody knows what it is. We quantify it with the use of regular or repetitive cycles only. Nobody can ever know what it is, but I am certain it is part of the cosmic setup that 'causes' things to come into being. Assuming this theory is true, then since everything biological is built upon a gene's prescriptions, time is written into all Beings; and the Beings in turn instinctively and unavoidable (because that is what it consists of) impose the regime on all other sub-beings or entities. Beings have to impose their basic natures on other things because they can't help it. This means the whole of human creations is attributable to the sense of time without which nothing could ever have been created, and it is so because time is cosmically the cause of the creation of man's basic nature.

This may be called "The logic of time in the universe", in the sense that every 'Being' in the cosmos will face the same problem and could only solve it in the same manner. The point is, having established this 'great' truth about time and life (since they go together so that it is impossible to speak about one and not the other), there is no longer any need to listen to religious leaders and slaughter other people in the name of any God because of the mystery surrounding time and life. With the knowledge we have of time as a secular entity, the life it is so closely associated with cannot have had divine origins, since scientifically the word divinity has no meaning.

Samuel K. K. Blankson

As we go to press, a nugget of wisdom against wars has come my way from a mere newspaper, demonstrating widespread liberal sentiments since Karl Popper, Russell and Einstein---perhaps we must add the influence of Professor Dixon's Human Situation as well: "The reality of the trenches, a war of mud, blood, cold, anger, boredom and the stench of men soiling themselves in abject fear, utterly overwhelmed the elevated rhetoric of duty and honour...The chief cultural consequences of the Great War...was a legacy of pervasive irony, a realism bred in the trenches among men who had become deaf to the ringing appeals of Victorian patriotism and the absurd demands of authority...At first it was fun. And then all of a sudden one realised what the infantry was for: it was for killing the maximum number of young men like you."[284] My advice is that nobody should rush to congratulate the men at CERN on the Higgs boson; for once 'the glue' that holds atoms and sub-atomic particles together to give mass is known, it will be used for wars---i.e. to make things, even mountains, and people disappear without trace.

---

[284] By Ben Macintyre, The London Times, 'Opinion', page 27, 15th June, 2012.

NOTES AND REFERENCES

Many sources are given in the text. The following are among the major books and papers on the subject known to me. Readers who are familiar with my work will notice that these references or some of them are used over and over again. The reason is that my theory of time is always the same, no matter how many books I write to try to elucidate it for commercial and literary reasons.

Time calls for the deepest thought about nature and life, but not much in the form of postulates to require massive tomes and intricate references in mathematics, logic and philosophy. It is so familiar that all that is required is an explanation why it is what it is. And being so difficult and mysterious there are not that many thinkers who have considered it in depth. For this reason my references for the many books on time are almost always the same.

ALBERT EINSTEIN (1879-1955)---SPACE TIME, an article in the 1926/27 (13th) edition of the Encyclopaedia Britannica. Also, RELATIVITY, in the same edition.

--------NATURE No. 106, 782, (1921), almost the whole issue was devoted to the confirmation of Einstein's new theory of gravity.

--------The Meaning of Relativity, Princeton University Press, 1966.

--------The Evolution of Physics, (With Leopold Infeld) Cambridge 1838.

--------RELATIVITY, Routledge Classics, London and New York, 2001.

HERMANN MINKOWSKI (1864-1909)----He first mentioned his supposition in a lecture in cologne, known as Raum und Zeit (Space and Time) Cologne 21st September, 1908.

--------Herman Minkowski AdP 47, 927 (1915)

--------Herman Minkowski, Goett. Nachr., 1908 p53. Reprinted in Gesammelte Abhandlungen von Herman Minkowski. Vol. 2, p352. Teubner, Leipzig 1911.

BERTRAND RUSSELL, FRS (1872-1970)----Our Knowledge of the External World, George Allen & Unwin, 1922.

--------Mysticism & Logic, George Allen & Unwin, 1976: a collection of important essays first published in 1917.

--------ABC OF RELATIVITY, George Allen & Unwin, 1958 (recently revised by Professor Felix Pirani---first published in 1925.

--------History of Western Philosophy, George Allen & Unwin, 1946.

--------My Philosophical Development, George Allen & Unwin, 1958.

--------The Analysis of Matter, George Allen & Unwin, 1927.

MORRIS KLINE: Mathematics in Western Culture, Allen & Unwin, London, 1954.

SIR ARTHUR STANLEY EDDINGTON, FRS (1862-1944)

--------The Expanding universe, University of Michigan Press, Ann Arbor, 1933

--------The Combination of Relativity Theory and Quantum Theory, Communication of the Dublin Institute of Advanced Studies, Dublin, 1943.

--------The Mathematical Theory of Relativity, Cambridge, second ed. 1930.

--------The Nature of the Physical World, Ann Arbor, Michigan, 1958.

--------Philosophy of Physical Science, Cambridge, 1949.

--------The Theory of Relativity and its Influence on Scientific Thought, Oxford, 1922.

--------Space, Time and Gravitation, Cambridge, 1920.

SIR JAMES JEANS, FRS: Physics and Philosophy, Cambridge, 1942.

--------The Mysterious Universe, Cambridge, 1930.

--------The New Background of Science, Cambridge, 1933.

PROFESSOR A.N. WHITEHEAD: The Concept of Nature, Ann Arbor, Michigan, 1957.

--------Science and the Modern World, Cambridge, 1922.

--------An Inquiry Concerning the Principle of Natural Knowledge, Cambridge, 1919.

--------Nature and Life, Cambridge, 1934.

--------Process and Reality: An Essay in Cosmology, Cambridge, 1929.

--------Essays in Science and Philosophy, Rider & Co., London, 1948.

--------The Principle of Relativity, Cambridge, 1922.

Professor BANESH HOFFMANN: The strange Story of the Quantum, Dover Pub. Inc. New York, 1959.

Professor STEVEN F. SAVITT (ed.) Times Arrows Today: Recent Physical and Philosophical Work on the Direction of Time, Cambridge, 1995.

CHARLES A. FRITZ: Bertrand Russell's Construction of the External World, Routledge & Kegan Paul, London, 1952.

Professor JEREMY BERNSTEIN: Albert Einstein and the Frontiers of Physics, Oxford, 1996.

Professor RICHARD FEYNMAN: Lectures---The Character of Physical law. MIT Press, 1967. There are several volumes of the Feynman lectures and they are all worthy of serious study.

Abraham Pais, "Subtle is The Lord: The Life and Science of Albert Einstein", Oxford, 1982. Professor Pais has methodically provided details of almost all the original papers relevant to relativity. His list is so exhaustive I don't know of a better one anywhere.

www.ingramcontent.com/pod-product-compliance
Lightning Source LLC
Chambersburg PA
CBHW020732180526
45163CB00001B/202